Lecture Notes
in Business Information Processing   315

Series Editors

Wil M. P. van der Aalst
RWTH Aachen University, Aachen, Germany
John Mylopoulos
University of Trento, Trento, Italy
Michael Rosemann
Queensland University of Technology, Brisbane, QLD, Australia
Michael J. Shaw
University of Illinois, Urbana-Champaign, IL, USA
Clemens Szyperski
Microsoft Research, Redmond, WA, USA

More information about this series at http://www.springer.com/series/7911

Ye Chen · Gregory Kersten
Rudolf Vetschera · Haiyan Xu (Eds.)

Group Decision and Negotiation in an Uncertain World

18th International Conference, GDN 2018
Nanjing, China, June 9–13, 2018
Proceedings

Editors
Ye Chen
Nanjing University of Aeronautics
    and Astronautics
Nanjing
China

Gregory Kersten ⓘ
Concordia University
Montreal, QC
Canada

Rudolf Vetschera ⓘ
University of Vienna
Vienna
Austria

Haiyan Xu ⓘ
Nanjing University of Aeronautics
    and Astronautics
Nanjing, Jiangsu
China

ISSN 1865-1348   ISSN 1865-1356 (electronic)
Lecture Notes in Business Information Processing
ISBN 978-3-319-92873-9   ISBN 978-3-319-92874-6 (eBook)
https://doi.org/10.1007/978-3-319-92874-6

Library of Congress Control Number: 2018946634

Printed on acid-free paper

This Springer imprint is published by the registered company Springer International Publishing AG
part of Springer Nature
The registered company address is: Gewerbestrasse 11, 6330 Cham, Switzerland

Preface

The annual conferences on Group Decision and Negotiation have become an important meeting point for researchers interested in the many aspects of collective decision-making. What started out as a one-time event at the beginning of the millennium has developed into a series of conferences that have been held (with one exception) every year since 2000. GDN is a truly global conference uniting researchers from all over the world, which to date has been held in four continents: Once each in Australia (Perth 2002) and South America (Recife 2012), four times in North America (Banff 2004, Mt. Tremblant 2007 and Toronto 2009, all in Canada, and Bellingham, USA 2016), and 11 times in Europe (Glasgow 2000, La Rochelle 2001, Istanbul 2003, Vienna 2005, Karlsruhe 2006, Coimbra 2008, Delft 2010, Stockholm 2013, Toulouse 2014, Warsaw 2015, and Stuttgart 2017).

GDN in 2018 came to Asia for the first time, and we are very grateful to the Nanjing University of Aeronautics and Astronautics for hosting this conference. In total, 143 papers grouped into 23 different streams were submitted for the conference, covering a wide range of topics related to group decisions and negotiations. Particularly large streams were, for example, conflict resolution (23 papers), preference modeling in GDN (13 papers), complex systems and decision analysis (13 papers), and consensus processes in decision-making (11 papers).

Out of these 143 papers, 15 papers were selected for inclusion in this volume based on a thorough review process. We have grouped these papers into three main areas: theoretical concepts of GDN, decision support and behavior in GDN, and applications of GDN.

The section on theoretical concepts of GDN contains three papers that present formal models intended to support groups of decision makers in various tasks. Xiaomei Mi and Huchang Liao in their paper "Hesitant Fuzzy Linguistic Group Decision-Making with Borda Rule" combine an innovative approach from soft computing, hesitant fuzzy sets, with one of the oldest methods of social choice, the Borda rule, to obtain a flexible and powerful tool of group decision-making. An important aspect in this kind of decision is data uncertainty, which is at the focus of the paper "A Multistage Risk Decision-Making Method for Normal Cloud Model Considering Three Reference Points" by Wen Song and Jianjun Zhu. Finally, Zhexuan Zhou, Xiangqian Xu, Yajie Dou, Yuejin Tan, and Jiang Jiang address this issue in the specific context of the selection of multiple decision alternatives in a portfolio setting in their paper "System Portfolio Selection Under Hesitant Fuzzy Information."

The papers in the second section deal with empirical studies of various methods to support groups and negotiators in their decision-making processes, and the behavioral effects that such support might have. In their paper "Representative Decision-Making and the Propensity to Use Round and Sharp Numbers in Preference Specification," Gregory E. Kersten, Ewa Roszkowska, and Tomasz Wachowicz look at an interesting bias phenomenon that might occur when supporting negotiations. Negotiation support models need a description of the negotiator's preferences; however, the elicitation process for these preferences could be distorted because users tend to provide "easy" numbers rather than exact information. The paper by Lucia Reis Peixoto Roselli, Eduarda Asfora Frej, and Adiel Teixeira de Almeida on "Neuroscience Experiment for Graphical Visualization in the FITradeoff Decision Support System" represents a new development in behavioral research on decision-making: Rather than just observing decisions and behavior, researchers begin more and more to measure the physiological processes during decision-making. As the authors show in their paper, physical data can indeed be used to, for example, better measure the cognitive effort in the decision process. Often, negotiators bargain on behalf of some institution or a principal, and then it is important that they are able to understand and follow the preferences of the group they are representing. The paper "On the Impact of the Negotiators' Culture, Background, and Instructions on the Representative Negotiation Process and Outcomes" by Tomasz Wachowicz, Gregory E. Kersten, and Ewa Roszkowska studies whether negotiators' ability to follow such instructions is dependent on their culture. The last two papers in this section study behavioral effects and preferences of negotiators in different application settings. Marta Dell'ovo, Eduarda Frej, Alessandra Oppio, Stefano Capolongo, Danielle Morais, and Adiel de Almeida describe the complex elicitation of preferences in a real life application in their paper "FITradeoff Method for the Location of Health-Care Facilities Based on Multiple Stakeholders' Preferences." Parmjit Kaur and Ashley Carreras in their paper "Capturing the Participants' Voice: Using Causal Mapping Supported by Group Decision Software to Enhance Procedural Justice" focus on an earlier phase of the decision process, the conceptual modeling of the decision problem, and study whether usual approaches to this modeling are fair in the sense that viewpoints of all participants are adequately taken into account. In the final paper of this section, Rustam Vahidov deals with an interesting way of influencing negotiator behavior. The paper shows that providing different images for an artificial opponent in human–computer negotiations will lead to different behavior on the side of the human negotiator.

The last section combines papers on specific application areas of group decisions and negotiations. The first two papers in this section are related to China's new silk road initiative. Shawei He, Ekaterina Flegentova, and Bing Zhu in their paper "Analyzing Conflicts of Implementing High-speed Railway Project in Central Asia Using Graph Model" study potential conflicts that might arise in creating the necessary traffic infrastructure, and how GDN tools like the graph model of conflict resolution might help to overcome them. A similar question is addressed in the paper "Strategic Negotiation for Resolving Infrastructure Development Disputes in One Belt One Road Initiative" by Waqas Ahmed, Qingmei Tan, and Sharafat Ali. The graph model is also applied to another global conflict in the paper "Attitudinal Analysis of Russia-Turkey Conflict with Chinese Role as a Third-Party Intervention" by Sharafat Ali and Haiyan Xu. The paper "Behavioral Modeling of Attackers Based on Prospect Theory and Corresponding Defenders Strategy" by Ziyi Chen, Chunqi Wan, Bingfeng Ge, Yajie Dou, and Yuejin Tan combines game theoretic analysis and the behavioral concepts of prospect theory to provide a more realistic model of a conflict situation. Last, but not least, the paper "A Multi-Stakeholder Approach for Energy Transition Policy Formation in Jordan" by Mats Danielson, Love Ekenberg, and Nadejda

Komendatova deals with conflicts and group decisions in the energy transition that many countries have to undergo.

The preparation of the conference and of this volume required the efforts and collaboration of many people. In particular, we thank the general chair of GDN 2018, Gregory Kersten, for his continuous support and effort that helped us to select the right papers for this volume and to carry out the review process within a short time. Special thanks also go to all the reviewers for their timely and informative reviews: Irene Abi-Zeid, Meng Chen, Shuding Chen, Ana Paula Costa, Suzana Daher, Xiao Deng, Luis Dias, Qingxing Dong, Love Ekenberg, Mohammad Feylizadeh, Michael Filzmoser, Amanda Garcia, Dorota Górecka, Masahide Horita, Zihan Jiang, Ginger Ke, Mark Kersten, Sabine Koeszegi, Tobias Langenegger, Jichao Li, Kevin Li, Xiaoning Lu, Philipp Melzer, Rafał Mierzwiak, Danielle Morais, Iván Palomares, Leandro C. Rego, Ewa Roszkowska, Yinxiaohe Sun, Liangyan Tao, Adiel Teixeira de Almeida, Ofir Turel, Tomasz Wachowicz, Junjie Wang, Liangpeng Wu, Yi Xiao, Fang Yinhai, Bo Yu, Pascale Zaraté, Hengjie Zhang, Shinan Zhao, and Jinhua Zhou.

We also are very grateful to Ralf Gerstner, Alfred Hofmann, and Christine Reiss at Springer publishers for the excellent collaboration.

May 2018

Ye Chen
Gregory Kersten
Rudolf Vetschera
Haiyan Xu

Organization

Honorary Chair

Marc Kilgour   Wilfrid Laurier University, Canada

General Chairs

Gregory Kersten   Concordia University, Montreal, Canada
Hong Nie   Nanjing University of Aeronautics and Astronautics, China

Program Chairs

Ye Chen   Nanjing University of Aeronautics and Astronautics, China
Rudolf Vetschera   University of Vienna, Austria
Haiyan Xu   Nanjing University of Aeronautics and Astronautics, China

Organizing Chairs

Keith Hipel   University of Waterloo, Canada
Dequn Zhou   Nanjing University of Aeronautics and Astronautics, China
Peng Zhou   Nanjing University of Aeronautics and Astronautics, China

Program Committee

Fran Ackermann	Curtin University, Australia
Adiel Almeida	Federal University of Pernambuco, Brazil
Deepinder Bajwa	Western Washington University, USA
Martin Bichler	Technical University of Munich, Germany
Xusen Cheng	University of International Business, China
João C. Clímaco	University of Coimbra, Portugal
Suzana F. D. Daher	Federal University of Pernambuco, Brazil
Colin Eden	Strathclyde University, Glasgow, UK
Gert-Jan de Vrede	University of Nebraska-Omaha, USA
Luis C. Dias	University of Coimbra, Portugal
Liping Fang	Ryerson University, Canada
Raimo Pertti Hämäläinen	Aalto University, Finland
Marlene Harlan	Western Washington University, USA
Masahide Horita	University of Tokyo, Japan
Takayuki Ito	Nagoya Institute of Technology, Japan
Catholjin Jonker	Delft University of Technology, The Netherlands
Sabine Koeszegi	Vienna University of Technology, Austria

Kevin Li	University of Windsor, Canada
Bilyana Martinovski	Stockholm University, Sweden
Paul Meerts	Clingendael Institute, The Netherlands
Ugo Merlone	University of Turin, Italy
Danielle Morais	Federal University of Pernambuco, Brazil
José Maria Moreno-Jiménez	Zaragoza University, Spain
Hannu Nurmi	University of Turku, Finland
Amer Obeidi	University of Waterloo, Canada
Pierpaolo Pontradolfo	Politecnico di Bari, Italy
Ewa Roszkowska	University of Białystok, Poland
Mareike Schoop	Hohenheim University, Germany
Melvin Shakun	New York University, USA
Wei Shang	Chinese Academy of Sciences, China
Rangaraja Sundraraj	Indian Institute of Management, India
Katia Sycara	Carnegie Mellon University, USA
Tomasz Szapiro	Warsaw School of Economics, Poland
Przemyslaw Szufel	Warsaw School of Economics, Poland
David P. Tegarden	Virginia Tech, USA
Ernest M. Thiessen	SmartSettle, Canada
Ofir Turel	California State University, USA
Rustam Vahidov	Concordia University, Canada
Doug Vogel	Harbin Institute of Technology, China
Tomasz Wachowicz	University of Economics Katowice, Poland
Christof Weinhardt	Karlsruhe Institute of Technology, Germany
Shi Kui Wu	University of Windsor, Canada
Yufei Yuan	McMaster University, Canada
Bo Yu	Concordia University, Canada
Pascale Zaraté	University of Toulouse 1, France
John Zeleznikow	Victoria University, Australia

Organizing Committee

Jian Chen	Nanjing University of Aeronautics and Astronautics, China
Gaofeng Da	Nanjing University of Aeronautics and Astronautics, China
Yeqing Guan	Nanjing University of Aeronautics and Astronautics, China
Shawei He	Nanjing University of Aeronautics and Astronautics, China
Xinjia Jiang	Nanjing University of Aeronautics and Astronautics, China
Linhan Ouyang	Nanjing University of Aeronautics and Astronautics, China
Zhitao Xu	Nanjing University of Aeronautics and Astronautics, China
Zhipeng Zhou	Nanjing University of Aeronautics and Astronautics, China

Contents

Theoretical Concepts of Group Decision and Negotiation

Hesitant Fuzzy Linguistic Group Decision Making with Borda Rule   3
   Xiaomei Mi and Huchang Liao

A Multistage Risk Decision Making Method for Normal Cloud Model
with Three Reference Points.   14
   Wen Song and Jianjun Zhu

System Portfolio Selection Under Hesitant Fuzzy Information.   33
   Zhexuan Zhou, Xiangqian Xu, Yajie Dou, Yuejin Tan, and Jiang Jiang

Decision Support and Behavior in Group Decision and Negotiation

Representative Decision-Making and the Propensity to Use Round
and Sharp Numbers in Preference Specification   43
   Gregory E. Kersten, Ewa Roszkowska, and Tomasz Wachowicz

Neuroscience Experiment for Graphical Visualization in the FITradeoff
Decision Support System.   56
   Lucia Reis Peixoto Roselli, Eduarda Asfora Frej,
   and Adiel Teixeira de Almeida

Impact of Negotiators' Predispositions on Their Efforts and Outcomes
in Bilateral Online Negotiations   70
   Bo Yu and Gregory E. Kersten

Some Methodological Considerations for the Organization and Analysis
of Inter- and Intra-cultural Negotiation Experiments.   82
   Tomasz Wachowicz, Gregory E. Kersten, and Ewa Roszkowska

FITradeoff Method for the Location of Healthcare Facilities Based
on Multiple Stakeholders' Preferences   97
   Marta Dell'Ovo, Eduarda Asfora Frej, Alessandra Oppio,
   Stefano Capolongo, Danielle Costa Morais,
   and Adiel Teixeira de Almeida

Capturing the Participants' Voice: Using Causal Mapping Supported
by Group Decision Software to Enhance Procedural Justice 113
   Parmjit Kaur and Ashley L. Carreras

The Effects of Photographic Images on Agent to Human Negotiations:
The Case of the Sicilian Clan 127
Rustam Vahidov

Applications of Group Decision and Negotiations

Analyzing Conflicts of Implementing High-Speed Railway Project
in Central Asia Using Graph Model 139
Shawei He, Ekaterina Flegentova, and Bing Zhu

Strategic Negotiation for Resolving Infrastructure Development Disputes
in the Belt and Road Initiative 154
Waqas Ahmed, Qingmei Tan, and Sharafat Ali

Attitudinal Analysis of Russia-Turkey Conflict with Chinese Role
as a Third-Party Intervention 167
Sharafat Ali, Haiyan Xu, Peng Xu, and Michelle Theodora

Behavioral Modeling of Attackers Based on Prospect Theory
and Corresponding Defenders Strategy........................ 179
Ziyi Chen, Chunqi Wan, Bingfeng Ge, Yajie Dou, and Yuejin Tan

A Multi-stakeholder Approach to Energy Transition Policy Formation
in Jordan 190
Mats Danielson, Love Ekenberg, and Nadejda Komendatova

Author Index 203

Theoretical Concepts of Group Decision and Negotiation

Hesitant Fuzzy Linguistic Group Decision Making with Borda Rule

Xiaomei Mi and Huchang Liao[(✉)]

Business School, Sichuan University, Chengdu 610064, China
mixiaomei2017@163.com, liaohuchang@163.com

Abstract. Hesitant fuzzy linguistic term set is an efficient tool to represent human thinking in decision making process. Borda rule is a powerful approach to aggregate opinions of a group members and it has been extended to several versions. In this paper, we investigate the Borda rule in the hesitant fuzzy linguistic context from both broad and narrow perspectives which are based on the possibility degree and the score function of the hesitant fuzzy linguistic term set. Moreover, we take into account the confidence level of linguistic evaluations with a parameter being generated from the evaluations. Finally, a numerical example is given to illustrate the efficiency of the Borda rule in group decision making within hesitant fuzzy linguistic information.

Keywords: Group decision making · Borda rule
Hesitant fuzzy linguistic term set · Confidence level

1 Introduction

Group decision making is a process in which the collective wisdom is used for the purpose of making a decision. Linguistic variables, first proposed by Zadeh [1], play a key role in describing human thoughts in group decision making process. For example, when judging the age of a person, linguistic terms such as *"young"*, *"middle-aged"* or *"old"* are less specific but more flexible than numerical values (see Fig. 1). Under some conditions, using only simple linguistic terms is not enough to depict the opinions of decision-makers. In this case, decision-makers cannot give accurate or precise term(s) for the alternatives but can give their evaluations using linguistic expressions such as *"at least young"*, *"between young and very young"* [2].

To fill this gap, the Hesitant Fuzzy Linguistic Term Set (HFLTS) was proposed in Ref. [3]. HFLTS allows the hesitation over several linguistic terms, which is much more powerful in expressing humans' ideas than simple and singleton linguistic term. It is much more applicable than traditional fuzzy linguistic approach and thus becomes to be an interesting research topic [4–6].

Borda rule was first proposed by Borda [7] for the social choice problems in the process of developing welfare economics in western countries. Considering the uncertainty of group decision making process, Borda rule has been extended to several circumstances, such as the fuzzy Borda rule [8], the hesitant fuzzy Borda rule [9] and the linguistic Borda rule [10]. García-Lapresta et al. [11] proposed the broad Borda rule and the narrow Borda rule in linguistic environment, which are different from the

© Springer International Publishing AG, part of Springer Nature 2018
Y. Chen et al. (Eds.): GDN 2018, LNBIP 315, pp. 3–13, 2018.
https://doi.org/10.1007/978-3-319-92874-6_1

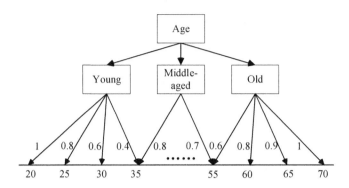

Fig. 1. The fuzzy linguistic variable "Age"

original Borda rule, as they transformed the linguistic evaluations into trapezoidal fuzzy numbers for the simplicity of comparison.

However, such a transformation may lose original information. By contrast, the HFLTS can capture principal concepts and ideas of decision-makers. As far as we know, there is little research with respect to the Borda rule [7] with hesitant fuzzy linguistic information. To fill these gaps, we try to propose the Borda rules with hesitant fuzzy linguistic information for group decision making. The Borda rule with hesitant fuzzy linguistic information is designed to tackle the uncertainty, fuzziness and indeterminacy existing in group decision making process, which broadens the application area of traditional Borda rule. The contributions of this proposal can be summarized as follows:

(1) We investigate the Borda rule in the hesitant fuzzy linguistic environment and call it Hesitant Fuzzy Linguistic Borda (HFL-Borda) rule. Considering that the HFLTS is an efficient tool in expressing the uncertainty existing in decision-making process, the fusion of these two useful tools makes the Borda rule more applicable.

(2) We consider the HFL-Borda count based on the possibility degrees and the score functions of the Hesitant Fuzzy Linguistic Elements (HFLEs). We define a new possibility degree formula of HFLEs for better comparison.

(3) Motivated by the Borda rule in linguistic decision making context, we extend the HFL-Borda rule to broad and narrow aspects based on the scores of HFLEs.

(4) We consider the confidence level varying with the lengths of HFLEs. The confidence level is derived from the hesitant fuzzy linguistic evaluations directly rather than from another evaluation scale given by the decision-maker again. Thus, it is very efficient in deriving the final results.

The paper is organized as follows: Sect. 2 reviews the Borda rule and the HFLTS. Section 3 introduces the hesitant fuzzy linguistic Borda rule based on the possible degree and score function of the HFLEs, respectively. Section 4 illustrates the calculation process. The paper ends in Sect. 5 with concluding remarks.

2 Preliminaries

In this section, some preliminary knowledge about the Borda rule and the HFLTS are introduced to facilitate further discussion.

2.1 The Borda Rule

The Borda rule [7] is designed for the social choice problems. Suppose that there are n alternatives $\{A_1, A_2, \cdots, A_i, \cdots, A_n\}$ under evaluation by decision-makers. For each decision-maker, the preferences on alternatives shall be shown in their judgements. For each alternative, according to the judgements of decision-makers, the Borda count of each alternative could be obtained, and it has two forms, broad and narrow [10]. Supposing that alternative A_i gets the ϑth rank by a decision-maker, then, in the view of this decision-maker, the Borda count of alternative A_i is $n - \vartheta + 1$. The larger the value of the Borda count is, the more superior the alternative should be.

2.2 The Hesitant Fuzzy Linguistic Term Set

Let $S = \{s_0, s_1, \cdots, s_{\tau-1}, s_\tau\}$ be a set of Linguistic Term Set (LTS). Rodríguez et al. [3] first introduced the HFLTS to represent the hesitation over several linguistic terms. For better understanding, Liao et al. [4] put forward the mathematical definition of HFLTS as:

$$H = \{(x, h_S(x)) | x \in X\} \tag{1}$$

where $h_S(x) = \{s_{\varphi(l)}(x) | s_{\varphi(l)}(x) \in S; l = 1, 2, \cdots, L\}$ is a HFLE representing the possible linguistic terms of x. The score of a HFLE h_S is defined as [4]:

$$\varsigma(h_S) = \frac{1}{L} \sum_{l=1}^{L} s_{\varphi(l)} = s_{\frac{1}{L} \sum_{l=1}^{L} \varphi(l)} \tag{2}$$

where $\varphi(l)$ is the subscripts of the linguistic term $s_{\varphi(l)}$ and L is the number of all linguistic terms in h_S.

3 Hesitant Fuzzy Linguistic Borda Rule

In this section, the Borda rule is investigated in the hesitant fuzzy linguistic environment from broad and narrow perspectives. Firstly, we consider the narrow perspective on Borda rule based on the possibility degrees of HFLEs. Afterwards, we investigate the broad and narrow Borda rules motivated by the idea of Ref. [10] based on the score function of HFLEs. The individual and group HFL-Borda rules are given as well. In addition, we study the confidence level of each HFLE to represent the belief degree of each judgment.

3.1   The Hesitant Fuzzy Linguistic Borda Rule Based on Possibility Degree

The possibility degree of HFLE has many forms as overviewed in Ref. [2]. One representative formula is defined as follows [11]:

$$P_d(h_S^1 \geq h_S^2) = \frac{\|h_S^1 \geq h_S^2\|}{\|h_S^1 \geq h_S^2\| + \|h_S^2 \geq h_S^1\|} \tag{3}$$

where $\|h_S^1 \geq h_S^2\|$ is the number of terms satisfying $s_i \geq s_j$; $s_i \in h_S^1$ and $s_j \in h_S^2$.

It is not difficult to find that the denominator of Eq. (3) is greater than $\|h_S^1 > h_S^2\| + \|h_S^1 = h_S^2\| + \|h_S^1 < h_S^2\|$ in the case that there are some common linguistic term(s) between h_S^1 and h_S^2. Given that $\|h_S^1 > h_S^2\|$, $\|h_S^1 = h_S^2\|$ and $\|h_S^1 < h_S^2\|$ are the complete relationships between h_S^1 and h_S^2, motivated by Eq. (3), we can propose a new formula to represent the possibility degree of h_S^1 being greater than h_S^2 as follows:

$$P(h_S^1 > h_S^2) = \frac{\|h_S^1 > h_S^2\|}{\|h_S^1 > h_S^2\| + \|h_S^1 = h_S^2\| + \|h_S^1 < h_S^2\|} \tag{4}$$

Correspondingly, the possibility degree of h_S^1 being greater than or equal to h_S^2 can be calculated by

$$P(h_S^1 \geq h_S^2) = P(h_S^1 > h_S^2) + P(h_S^1 = h_S^2) = \frac{\|h_S^1 > h_S^2\| + \|h_S^1 = h_S^2\|}{\|h_S^1 > h_S^2\| + \|h_S^1 = h_S^2\| + \|h_S^1 < h_S^2\|}$$

We can find that $P(h_S^1 \geq h_S^2)$ is greater than $P_d(h_S^1 \geq h_S^2)$ in the case that there are some common linguistic term(s) between h_S^1 and h_S^2, because the denominator of $P(h_S^1 \geq h_S^2)$ is less than $P_d(h_S^1 \geq h_S^2)$. Moreover, the denominator of $P(h_S^1 \geq h_S^2)$ is more reasonable than $P_d(h_S^1 \geq h_S^2)$ as $P_d(h_S^1 \geq h_S^2)$ counts the equivalent terms twice.

Another possibility degree formula is defined as [3]

$$P(h_S^1 > h_S^2) = \frac{\max[0, \varphi(h_S^{1+}) - \varphi(h_S^{2-})] - \max[0, \varphi(h_S^{1-}) - \varphi(h_S^{2+})]}{[\varphi(h_S^{1+}) - \varphi(h_S^{1-})] + [\varphi(h_S^{2+}) - \varphi(h_S^{2-})]} \tag{5}$$

This formula is too simple and is not able to tackle the condition in which the same linguistic term(s) appears in two HFLEs. For instance, let $h_S^1 = \{s_1, s_2\}$ and $h_S^2 = \{s_2, s_3, s_4\}$, by Eq. (5), we obtain $P(h_S^1 > h_S^2) = 1, P(h_S^1 < h_S^2) = 1$. It is not difficult to find that the results calculated by Eq. (5) are unreasonable. While using Eq. (4), we have $P(h_S^1 > h_S^2) = 0, P(h_S^1 = h_S^2) = 1/6, P(h_S^1 < h_S^2) = 5/6$, which are more reasonable.

Compared with Eqs. (3) and (5), the proposed possibility degree formula, Eq. (4), shows advantages in comparing HFLEs. Therefore, in the following presentation, we use Eq. (4) to calculate the possibility degree of a HFLE being greater than the medium linguistic term to obtain the Borda count in hesitant fuzzy linguistic decision making context.

Suppose that there are n alternatives $\{A_1, A_2, \cdots, A_i, \cdots, A_n\}$ judged by m decision-makers $\{DM_1, DM_2, ..., DM_k, ..., DM_m\}$ over the LTS, $S = \{s_0, s_1, \cdots, s_{\tau-1}, s_\tau\}$. Considering the efficiency of HFLTS in expressing hesitant fuzzy linguistic opinions, we assume that the decision-makers give their judgements in pairwise comparisons over the alternatives with linguistic expressions. For example, suppose that decision-maker DM_k gives the linguistic expression $ll_S^{ij(k)}$ towards alternative A_i over alternative A_j. Based on the transformation function [3], the linguistic evaluations of the decision-makers DM_k on A_i over A_j can be transformed to the HFLE $h_S^{ij(k)}$. In this way, we can construct the individual linguistic evaluation matrix $LL_S^{(k)}$ and the individual hesitant fuzzy linguistic decision matrix $H_S^{(k)} = \left(h_S^{ij(k)} \right)_{n \times n}$ of the decision-maker DM_k.

$$
LL_S^{(k)} = \begin{array}{c} \\ A_1 \\ A_2 \\ \vdots \\ A_n \end{array} \overset{\begin{array}{cccc} A_1 & A_2 & \cdots & A_n \end{array}}{\left(\begin{array}{cccc} ll_S^{11(k)} & ll_S^{12(k)} & \cdots & ll_S^{1n(k)} \\ ll_S^{21(k)} & ll_S^{22(k)} & \cdots & ll_S^{2n(k)} \\ \vdots & \vdots & \cdots & \vdots \\ ll_S^{n1(k)} & ll_S^{n2(k)} & \cdots & ll_S^{nn(k)} \end{array} \right)}, \quad k = 1, 2, \cdots, m
$$

$$
H_S^{(k)} = \begin{array}{c} \\ A_1 \\ A_2 \\ \vdots \\ A_n \end{array} \overset{\begin{array}{cccc} A_1 & A_2 & \cdots & A_n \end{array}}{\left(\begin{array}{cccc} h_S^{11(k)} & h_S^{12(k)} & \cdots & h_S^{1n(k)} \\ h_S^{21(k)} & h_S^{22(k)} & \cdots & h_S^{2n(k)} \\ \vdots & \vdots & \cdots & \vdots \\ h_S^{n1(k)} & h_S^{n2(k)} & \cdots & h_S^{nn(k)} \end{array} \right)}, \quad k = 1, 2, \cdots, m
$$

To derive the group decision result, we introduce some definitions related to Borda rule under the hesitant fuzzy linguistic environment.

(1) The Hesitant Fuzzy Linguistic Individual Narrow Borda Count (HFL-INBC) is defined as

$$
HFL\text{-}INBC_1^{(k)}(A_i) = \sum_{j=1}^n P(h_S^{ij(k)} > s_{\tau/2}), k = 1, 2, \cdots, m \qquad (6)
$$

where $P(h_S^{ij(k)} > s_{\tau/2})$ represents the possibility degree of $h_S^{ij(k)}$ being greater than the medium linguistic term $s_{\tau/2}$.

(2) The Hesitant Fuzzy Linguistic Group Narrow Borda Count (HFL-GNBC) is defined as

$$
HFL\text{-}GNBC_1(A_i) = \sum_{k=1}^m HFL\text{-}INBC_1^{(k)}(A_i) \qquad (7)
$$

Example 1. There are two decision-makers DM_1 and DM_2 evaluating three alternatives A_1, A_2 and A_3 with HFLTSs. The evaluations of these decision-makers over three alternatives are shown as follows:

$$H_S^{(1)} = \begin{pmatrix} \{s_3\} & \{s_4, s_5, s_6\} & \{s_5, s_6\} \\ \{s_0, s_1, s_2\} & \{s_3\} & \{s_1, s_2, s_3\} \\ \{s_0, s_1\} & \{s_3, s_4, s_5\} & \{s_3\} \end{pmatrix}$$

$$H_S^{(2)} = \begin{pmatrix} \{s_3\} & \{s_2, s_3, s_4\} & \{s_4, s_5, s_6\} \\ \{s_2, s_3, s_4\} & \{s_3\} & \{s_3, s_4\} \\ \{s_0, s_1, s_2\} & \{s_2, s_3\} & \{s_3\} \end{pmatrix}$$

By Eq. (6), we can calculate the HFL-INBC of the alternatives related to DM_1 as $HFL-INBC_1^{(1)}(A_1) = 2$, $HFL-INBC_1^{(1)}(A_2) = 0$, $HFL-INBC_1^{(1)}(A_3) = 2/3$. The HFL-INBCs of the alternatives related to DM_2 are 4/3, 2/3 and 0, respectively. Furthermore, we can calculate the HFL-GNBCs of the alternatives by Eq. (7): $HFL-GNBC(A_1) = 10/3$, $HFL-GNBC(A_2) = 2/3$, $HFL-GNBC(A_3) = 2/3$. Thus, A_1 is superior to A_2 and A_3, but the tie happens between A_2 and A_3 even though the evaluations of two decision-makers over these two alternatives are different. This weakness is due to the fact that the HFL-INBC and the HFL-GNBC, which are based on the possibility degrees of the hesitant fuzzy linguistic judgments that are greater than the medium linguistic term, involve only the simple counts of the relations between the linguistic terms and the medium linguistic term, without considering the intensity of their relations. That is to say, the superiority or inferiority degree of each element over the medium linguistic term is not considered in the above Borda counts. To overcome this disadvantage, in the next subsection, another form of HFL-Borda rule considering the intensities of superiority and inferiority degrees is introduced.

3.2 The Hesitant Fuzzy Linguistic Borda Rule Based on Score Function

Motivated by the Borda rule in the linguistic decision making context [12], we can introduce some definitions of Borda count under the hesitant fuzzy linguistic environment based on the score function of HFLEs.

(1) The Hesitant Fuzzy Linguistic Individual Broad Borda Count (HFL-IBBC) of alternative A_i is defined as

$$HFL-BBC_2^k(A_i) = \frac{1}{n}\sum_{j=1}^n \varsigma(h_S^{ij(k)}) \tag{8}$$

The broad Borda count defined as Eq. (8) represents the overall relationship of alternative A_i over the others. It involves all the preference levels, including the superiority, the inferiority and indifference.

(2) The Hesitant Fuzzy Linguistic Individual Narrow Borda Count (HFL-INBC) of alternative A_i is defined as

$$HFL-INBC_2^k(A_i) = \frac{1}{q}\sum\nolimits_{j=1,\, \varsigma(h_S^{ij(k)}) > s_{\tau/2}}^{n} \varsigma(h_S^{ij(k)}) \qquad (9)$$

where q represent the number of alternatives that satisfy $\varsigma(h_S^{ij(k)}) > s_{\tau/2}$.

The narrow Borda count defined as Eq. (9) refers to the overall relationships of alternative A_i being greater than the others (whose scores regarding to A_i are greater than the medium term $s_{\tau/2}$, i.e., $\varsigma(h_S^{ij(k)}) > s_{\tau/2}$). We should note that $HFL-INBC_2^k(A_i)$ could be zero in the case where there is no term greater than $s_{\tau/2}$.

(3) The Hesitant Fuzzy Linguistic Group Broad Borda Count (HFL-GBBC) of alternative A_i is defined as

$$HFL-GBBC_2(A_i) = \frac{\sum_{k=1}^{m} w_k \times HFL-IBBC_2^k(A_i)}{\sum_{k=1}^{m} w_k} \qquad (10)$$

where w_k is the weight of the DM_k.

(4) The Hesitant Fuzzy Linguistic Group Narrow Borda Count (HFL-GNBC) of alternative A_i is defined as

$$HFL-GNBC_2(A_i) = \frac{\sum_{k=1}^{m} w_k \times HFL-INBC_2^k(A_i)}{\sum_{k=1}^{m} w_k} \qquad (11)$$

where w_k is the weight of the DM_k.

There are three ways to get the final rankings of alternatives. The first one only considers the HFL-IBBCs, the second one considers the HFL-INBCs, while the third one considers the HFL-GBBCs and the HFL-GNBCs of alternatives. The rankings derived from the HFL-IBBCs denote the whole degrees of superiority and inferiority over other alternatives, and the rankings derived from the HFL-INBCs only denote the superiority degrees over other alternatives. The third rankings with respect to the values of the HFL-GBBCs and the HFL-GNBCs of alternatives are the overall dominance ranking of the alternatives.

3.3 The Confidence Level of the HFLE

In the linguistic decision making context, the Borda rule has been extended by the confidence level of each evaluation value given by decision-makers (please see Refs. [12, 13] for details). The confidence level multiplying the given singleton linguistic term can be used to represent the real evaluation of the decision-maker. With the help of the confidence level, the cognitions of the decision-maker can be expressed much more comprehensively than only using the linguistic term. In this paper, we also use this way to represent the decision-makers' opinions and then aggregate the confidence levels and their associated evaluations. However, it is not difficult to find that two times evaluations over two variables wastes time and effort, especially when the number of alternatives is large. Therefore, we try to use a parameter, which varies with the length of the HFLE, to replace the confidence level, shown as:

$$\delta(h_S) = {}^{[\#S-\#(h_S)]}\!\!\sqrt{[\#S - \#(h_S)]/(\#S - 1)} \qquad (12)$$

where $\#S$ and $\#(h_S)$ are the lengths of the LTS S and the specific HFLE h_S, respectively.

The $\#S - \#(h_S)$ the root of the number $[\#S - \#(h_S)]/(\#S - 1)$ is the confidence level of HFLE h_S. Other functions in Ref. [14] can also be used for obtaining the confidence level, which is an interesting research topic for future work.

When $\#(h_S) = \#S$, the linguistic judgement given by decision-maker is the complete LTS. That is to say, the decision-makers could not formulate the results of their thinking process clearly and thus the confidence level $\delta(h_S) = 0$. While $\#(h_S) = 1$ implies that the decision-makers do not hesitate over the linguistic terms, so the linguistic judgement h_S should be considered fully without any discount and thus we let $h_S = 1$. The confidence levels related to the well-known seven-valued LTS are listed in Table 1.

Table 1. The confidence levels related to seven-valued LTS

$\#(h_S)$	1	2	3	4	5	6	7
$\delta(h_S)$	1	0.9642	0.9036	0.7937	0.5774	0.1667	0

Let $S = \{s_0, s_1, \cdots, s_5, s_6\}$ be a LTS and $h_S = \{s_4, s_5, s_6\}$. The confidence level of h_S is $\delta(h_s) = 0.9036$. With the confidence level, the score of h_S can be modified as $\varsigma'(h_S) = s_{\frac{1}{L}\sum_{l=1}^{L}\varphi(l)} * \delta(h_S) = s_{4.518}$. As $\varsigma'(h_S) < \varsigma(h_S)$, the multiple hesitant linguistic terms included in h_S weaken the confidence degree of the evaluation information.

4   Numerical Example

Suppose that three decision-makers DM_1, DM_2 and DM_3 give the evaluations of three alternatives A_1, A_2 and A_3 in linguistic expressions. The LTS S is the same as that in Example 1. The hesitant fuzzy linguistic decision matrices are shown as follows:

$$H_S^{(1)} = \begin{pmatrix} \{s_3\} & \{s_4, s_5, s_6\} & \{s_5, s_6\} \\ \{s_0, s_1, s_2\} & \{s_3\} & \{s_1, s_2, s_3\} \\ \{s_0, s_1\} & \{s_3, s_4, s_5\} & \{s_3\} \end{pmatrix}$$

$$H_S^{(2)} = \begin{pmatrix} \{s_3\} & \{s_2, s_3, s_4\} & \{s_4, s_5, s_6\} \\ \{s_2, s_3, s_4\} & \{s_3\} & \{s_3, s_4\} \\ \{s_0, s_1, s_2\} & \{s_2, s_3\} & \{s_3\} \end{pmatrix}$$

$$H_S^{(3)} = \begin{pmatrix} \{s_3\} & \{s_5, s_6\} & \{s_0, s_1, s_2\} \\ \{s_0, s_1\} & \{s_3\} & \{s_6\} \\ \{s_4, s_5, s_6\} & \{s_0\} & \{s_3\} \end{pmatrix}$$

The HFL-IBBCs and the HFL-INBCs of the decision-maker DM_3 can be calculated by Eqs. (8) and (9) as follows:

$$HFL{-}IBBC_2^{(3)}(A_1) = s_{3.17}, HFL{-}IBBC_2^{(3)}(A_2) = s_{3.17}, HFL{-}IBBC_2^{(3)}(A_3) = s_{2.67}.$$

$$HFL{-}INBC_2^{(3)}(A_1) = s_{5.5}, HFL{-}INBC_2^{(3)}(A_2) = s_6 HFL{-}INBC_2^{(3)}(A_3) = s_5.$$

The different Borda counts introduced in Sect. 3 can be calculated by Eqs. (8)–(11) and tabulated in Table 2.

Table 2. The different Borda counts and ranks of the alternatives

	DM_1				DM_2				DM_3				Group			
	$B^{(1)}$	R	$N^{(1)}$	R	$B^{(2)}$	R	$N^{(2)}$	R	$B^{(3)}$	R	$N^{(3)}$	R	B	R	N	R
A_1	$s_{4.5}$	1	$s_{5.25}$	1	$s_{3.67}$	1	$s_{4.5}$	1	$s_{3.17}$	1	$s_{5.5}$	2	$s_{3.78}$	1	$s_{5.08}$	1
A_2	s_2	3	s_3	3	$s_{3.17}$	2	$s_{3.5}$	2	$s_{3.17}$	1	s_6	1	$s_{2.78}$	2	$s_{4.17}$	2
A_3	$s_{2.5}$	2	$s_{4.5}$	2	$s_{2.17}$	3	s_3	3	$s_{2.67}$	3	s_5	3	$s_{2.45}$	3	$s_{4.17}$	2

Note. $B^{(k)}$ and $N^{(k)}$ refer to the HFL-IBBCs and HFL-INBCs of alternatives, respectively. R means the ranking of alternatives in different situations.

It is not hard to figure out that the group prefers A_1 to A_2, and prefers A_2 to A_3 from the results of the HFL-GBBCs. But the results of the HFL-GNBCs, show tie between A_2 and A_3. Considering these two Borda counts together, the ranking $A_1 > A_2 > A_3$ can be derived. Moreover, the results considering confidence levels are shown in Table 3.

Table 3. The different Borda counts with confidence levels and ranks of the alternatives

	DM_1				DM_2				DM_3				Group			
	$B'^{(1)}$	R	$N'^{(1)}$	R	$B'^{(2)}$	R	$N'^{(2)}$	R	$B'^{(3)}$	R	$N'^{(3)}$	R	B'	R	N'	R
A_1	$s_{4.1626}$	1	$s_{4.7439}$	1	$s_{3.4096}$	1	$s_{4.518}$	1	$s_{3.0689}$	1	$s_{5.3031}$	2	$s_{3.5741}$	1	$s_{4.855}$	1
A_2	$s_{1.9036}$	3	s_3	3	$s_{3.0285}$	2	$s_{3.3747}$	2	$s_{3.1607}$	2	s_6	1	$s_{2.6976}$	2	$s_{4.6874}$	2
A_3	$s_{2.3655}$	2	$s_{3.6144}$	2	$s_{2.1047}$	3	s_3	3	$s_{2.506}$	3	$s_{4.518}$	3	$s_{2.3254}$	3	$s_{3.8775}$	3

Note. $B'^{(k)}$ and $N'^{(k)}$ are the HFL-IBBCs and HFL-INBCs of alternatives with the confidence level, respectively. R means the ranking of alternatives on different situations.

Comparing with the results in Table 3, the same ranking $A_1 > A_2 > A_3$ is obtained from the values of HFL-GBBCs and HFL-GNBCs of alternatives without ties. It is not hard to find that the confidence level plays a prominent role in reducing the ties.

The same evaluation was provided by the second decision maker $H_S^{(2)}$ in Example 1 and this numerical example. However, in Example 1, A_2 and A_3 cannot be distinguished because of lacking the no-confidence level combined Borda rule. Hence, it is necessary to use the confidence level which contributes to breaking ties.

5 Conclusion

This paper proposed the Borda rule in the hesitant fuzzy linguistic decision making context from the broad and narrow perspectives. To calculate the Borda counts, two different ways, possibility degree based method and score function based method, were adopted. Considering the flexibility of HFLTSs, the confidence level was taken into consideration, which varies with the lengths of evaluations given by decision-makers.

In the future, two-dimensional work should be considered. From the practical perspective, hesitant fuzzy linguistic group decision making with Borda rule can be used in voting for the chairman or other elections. Different functions [14] to obtain confidence level of linguistic judgements can be compared to get an appropriate function in group decision making. From the theory level, group decision making with Borda rule can also be extended into the probabilistic linguistic environment [15].

Acknowledgements. The work was supported by the National Natural Science Foundation of China (71501135, 71771156), the 2016 Key Project of the Key Research Institute of Humanities and Social Sciences in Sichuan Province (CJZ16-01, CJCB2016-02, Xq16B04), and the Scientific Research Foundation for Excellent Young Scholars at Sichuan University (No. 2016SCU04A23).

References

1. Zadeh, L.A.: The concept of a linguistic variable and its application to approximate reasoning. Inf. Sci. **8**(3), 199–249 (1975)
2. Liao, H.C., Xu, Z.S., Enrique, H.V., Herrera, F.: Hesitant fuzzy linguistic term set and its application in decision making: A state of the art survey. Int. J. Fuzzy Syst. (2018, in press). https://doi.org/10.1007/s40815-017-0432-9
3. Rodriguez, R.M., Martinez, L., Herrera, F.: Hesitant fuzzy linguistic term sets for decision making. IEEE Trans. Fuzzy Syst. **20**(1), 109–119 (2012)
4. Liao, H.C., Xu, Z.S., Zeng, X.J., Merigó, J.M.: Qualitative decision making with correlation coefficients of hesitant fuzzy linguistic term sets. Knowl. Based Syst. **76**, 127–138 (2015)
5. Liao, H.C., Xu, Z.S., Zeng, X.J.: Hesitant fuzzy linguistic VIKOR method and its application in qualitative multiple criteria decision making. IEEE Trans. Fuzzy Syst. **23**(5), 1343–1355 (2015)
6. Liao, H.C., Yang, L.Y., Xu, Z.S.: Two new approaches based on ELECTRE II to solve the multiple criteria decision making problems with hesitant fuzzy linguistic term sets. Appl. Soft Comput. **63**, 223–234 (2018)
7. Borda, J.C.: Mémoire sur les Élections au Scrutin, Histoire de l, Académie Royale des Sciences. Paris, France (1781)

8. García-Lapresta, J.L., Martínez-Panero, M.: Borda count versus approval voting: a fuzzy approach. Public Choice **112**(1), 167–184 (2002)
9. Khalid, A., Beg, I.: Incomplete hesitant fuzzy preference relations in group decision making. Int. J. Fuzzy Syst. **19**(3), 637–645 (2017)
10. García-Lapresta, J.L., Martínez-Panero, M., Meneses, L.C.: Defining the Borda count in a linguistic decision making context. Inf. Sci. **179**(14), 2309–2316 (2009)
11. Hesamian, G., Shams, M.: Measuring similarity and ordering based on hesitant fuzzy linguistic term sets. J. Intell. Fuzzy Syst. **28**, 983–990 (2015)
12. Zeyuan, Q., Dosskey, M.G., Kang, Y.: Choosing between alternative placement strategies for conservation buffers using Borda count. Landscape Urban Plann. **153**, 66–73 (2016)
13. Jin, Z., Qiu, Z.: An improved fuzzy Borda count and its application to watershed management. In: International Conference on Renewable Energy and Environmental Technology (2017)
14. Harrison, C., Amento, B., Kuznetsov, S., Bell, R.: Rethinking the progress bar. In: ACM Symposium on User Interface Software and Technology, Newport, Rhode Island, USA, October, pp. 115–118. DBLP (2007)
15. Wu, X.L., Liao, H.C.: An approach to quality function deployment based on probabilistic linguistic term sets and ORESTE method for multi-expert multi-criteria decision making. Inf. Fusion **43**, 13–26 (2018)

A Multistage Risk Decision Making Method for Normal Cloud Model with Three Reference Points

Wen Song and Jianjun Zhu[(✉)]

College of Economics and Management,
Nanjing University of Aeronautics and Astronautics,
Nanjing 211106, People's Republic of China
{songwen, zhujianjun}@nuaa.edu.cn

Abstract. Decision making problems become more complicated due to the dynamically changing environment. Consequently, decision making methods with reference points are increasing. Reference points provide a good basis for decision makers. This paper proposes a multistage risk decision making method for normal cloud model considering three reference points. Firstly, the setting method of three reference points is proposed considering the dimensions of multistate, development and promotion. The value function is defined based on the characteristics of three reference points. Secondly, the aggregation methods for different prospect values are proposed with the preference coefficients, which are calculated by the synthetic degree of grey incidence. Thirdly, a two-stage weight optimization method is proposed to solve the attribute weights and stage weights based on the idea of minimax reference point optimization. Finally, a numerical example illustrates the feasibility and validity of the proposed method.

Keywords: Multistage risk decision making · Three reference points
Normal cloud model · Two-stage weight optimization method

1 Introduction

Multistage risk decision making (MSRDM) methods aim to rank alternatives or select the best alternative(s) by the aggregation of multistage risk decision-making (DM) information. MSRDM problems include risk, uncertainty and dynamics. Psychological factors of the decision-makers need to be taken into consideration to solve the risk DM problems. Decision-makers often consider the gain and loss under a reference point due to the bounded rationality. The fairness and satisfaction of DM is significantly influenced by a single reference point. In dynamic and uncertain conditions, using a single reference point will lead to the loss of some of the information about the distribution of the results. In the DM process with risk and dynamics, the psychological behavior of decision makers is inconsistent. In this context, the consideration of multiple reference points helps decision-makers to uncover the dynamic and risk characteristics of MSRDM problems, thus making a reasonable and comprehensive assessment of results.

© Springer International Publishing AG, part of Springer Nature 2018
Y. Chen et al. (Eds.): GDN 2018, LNBIP 315, pp. 14–32, 2018.
https://doi.org/10.1007/978-3-319-92874-6_2

Work on MSRDM methods have been increasing recently. In a multistage DM problem of finite-state automaton, a new optimization method stochastically develops a solution step-by-step in combination with a simulated annealing [1]. In an optimal investment problem with several projects, a new methodology is proposed based on experts' evaluations. It consists of three stages [2]. The multistage one-shot DM problems under uncertainty are studied based on scenario [3]. A multi-stage technical screening and evaluation tool is proposed to determine the optimal technique scheme under fuzzy environment [4]. A multistage assignment model is presented for rescue teams to dynamically respond to the disaster chain [5].

With the increasing complexity of DM problems, more effective methods are developed to support decision makers' judgments. DM methods considering reference points are one kind of resourceful methods. TOPSIS [6, 7] is widely used in MCDM problems. The idea of TOPSIS is to compare each solution with the positive ideal solution and the negative ideal solution, which are actually the two reference points. VIKOR methods [8] are also dependent on the positive and negative ideal solution, which are similar to the TOPSIS methods. Kahneman and Tversky [9, 10] presented the Prospect Theory to solve the risky DM problems.

The actual utility is obtained from comparison with a reference point. Due to the limited rationality of decision-makers' thinking, it is difficult to judge by the evaluation value. The consideration of reference points can provide the basis for decision makers, and lead to better informed and well-reasoned decisions. DM methods considering the reference point have been gradually enriched. A prospect theory-based interval dynamic reference point method has been proposed for emergency DM [11]. A risk DM method has been proposed considering the dynamic reference point, the external reference point and the internal reference point [12]. A new method based on the concept of ideal solution has been presented as a possible variant of TOPSIS and VIKOR methods [13]. The newsvendor's pricing and stocking decisions have been studied considering the impact of reference point effects [14].

Decision information often shows different forms, such as fuzzy numbers [2], interval numbers [12], linguistic sets [15], cloud models and so on. Due to the dynamic continuity of MSRDM process and the risky DM environment, information often shows fuzziness and randomness at the same time. The transformation between qualitative concepts and quantitative concepts is often needed to be dealt with. Linguistic set is usually used to express the decision maker's judgment. However, linguistic sets are often ambiguous and uncertain, and very difficult to form accurate information [15]. In this context, Li presented cloud models to propose conversion between qualitative concept and quantitative representation [16]. Many new approaches to cloud models have been proposed to solve existing problems. Cloud Hierarchical Analysis (CHA) is an extension of AHP [17]. The Cloud Delphi hierarchical analysis has been presented for practical multi-criteria group DM problems [18]. DM methods combining linguistic sets and cloud model have been investigated [19]. Cloud model has been widely used in many problems, such as water quality assessment [20], image segmentation [21], and clustering problems [22].

This paper makes contributes to the MSRDM problems for normal cloud model with three reference points.

(1) The new setting method for three reference points is proposed based on the data in different states and stages.
(2) The aggregation method for different prospect values is proposed based on the synthetic degree of grey incidence.
(3) A two-stage weight optimization method is proposed to solve the attribute weights and stage weights.

The rest of this paper is organized as follows. In Sect. 2, some related concepts and definitions are reviewed. Section 3 presents the MSRDM method for normal cloud model with three reference points. Section 4 provides a case followed by its analysis. Section 5 concludes the paper.

2 Preliminaries

This section briefly reviews the basic concepts and definitions associated with normal cloud model and prospect theory, and describes the problem addressed in this paper.

2.1 Basic Concepts and Definitions

Due to the complexity and uncertainty of DM problems, DM information often shows fuzziness and uncertainty. Normal cloud model provides an important background to represent fuzziness and randomness at the same time. Many scholars [18, 20, 23, 24] have carried out studies using normal cloud model.

Definition 1. [23] Let U be the universe of discourse and $\tilde{A}$ be a qualitative concept in U. If $x \in U$ is a random instantiation of concept $\tilde{A}$ that satisfies $x \sim N(Ex, En'^2)$, $En' \sim N(En, He^2)$, and the certainty degree of x belonging to $\tilde{A}$ satisfies

$$y = e^{-\frac{(x-Ex)^2}{2(En')^2}} \tag{1}$$

The distribution of x in the universe U is called the normal cloud model and x can be called a cloud drop. The normal cloud model can effectively integrate the randomness and fuzziness of a concept through three parameters: Expectation Ex, Entropy En and Hyper Entropy He. Expectation Ex is the mathematical expectation of the cloud drops belonging to a concept in the universe. It can best represent the qualitative concept. Entropy En represents the uncertainty measurement of a qualitative concept. It is the measurement of randomness and fuzziness of the concept. Hyper Entropy He is the uncertain degree of entropy En [23].

Given two normal cloud models $C_i(Ex_i, En_i, He_i)$ and $C_j(Ex_j, En_j, He_j)$. Certain operation rules between two normal cloud models have been included in [18, 20].

(1) $C_i + C_j = \left(Ex_i + Ex_j, \sqrt{En_i^2 + En_j^2}, \sqrt{He_i^2 + He^2} \right)$.

(2) $C_i - C_j = \left(Ex_i - Ex_j, \sqrt{En_i^2 + En_j^2}, \sqrt{He_i^2 + He^2} \right)$.

(3) $C_i \times C_j = \left(Ex_i \times Ex_j, |Ex_iEx_j|\sqrt{\left(\frac{En_i}{Ex_i}\right)^2 + \left(\frac{En_j}{Ex_j}\right)^2}, |Ex_iEx_j|\sqrt{\left(\frac{He_i}{Ex_i}\right)^2 + \left(\frac{He_j}{Ex_j}\right)^2} \right).$

(4) $C_i \div C_j = \left(\frac{Ex_i}{Ex_j}, \left|\frac{Ex_i}{Ex_j}\right|\sqrt{\left(\frac{En_i}{Ex_i}\right)^2 + \left(\frac{En_j}{Ex_j}\right)^2}, \left|\frac{Ex_i}{Ex_j}\right|\sqrt{\left(\frac{He_i}{Ex_i}\right)^2 + \left(\frac{He_j}{Ex_j}\right)^2} \right).$

(5) $\lambda C_i = \left(\lambda Ex_i, \sqrt{\lambda}En_i, \sqrt{\lambda}He_i \right).$

(6) $(C_i)^\lambda = \left(Ex_i^\lambda, \sqrt{\lambda}Ex_i^{\lambda-1}En_i, \sqrt{\lambda}Ex_i^{\lambda-1}He_i \right).$

When applying normal cloud model, the similarity between two normal cloud models is commonly used.

Definition 2. [24] Let $C_i(Ex_i, En_i, He_i)$ and $C_j(Ex_j, En_j, He_j)$ be two normal cloud models. The similarity between two normal cloud models based on shape and distance is defined as:

$$sim_c(C_i, C_j) = sim_d(C_i, C_j) \times sim_s(C_i, C_j) \tag{2}$$

Where $sim_d(C_i, C_j)$ represents the similarity between two normal cloud models based on distance, $sim_s(C_i, C_j)$ represents the similarity between two normal cloud models based on shape.

In prospect theory, alternatives are selected based on the prospect value.

Definition 3. [9] The prospect value is defined by the value function and the probability weight function:

$$V(x) = \sum_k \pi(p_k) \cdot v(x_k) \tag{3}$$

Definition 4. [10] The value function is expressed in the form of a power law according to the following formula:

$$v(x_k) = \begin{cases} (x_k)^\alpha, & x_k \geq 0 \\ -\theta(-x_k)^\beta & x_k < 0 \end{cases} \tag{4}$$

Where x_k denotes the gain or loss of the value when comparing an alternative to its reference point. When $x_k \geq 0$, it represents a gain. When $x_k < 0$, it represents a loss. α and β represent the concave and convex degree of the value power function $v(x_k)$ in the region of gain and loss respectively. θ indicates the loss-averse coefficient.

Definition 5. [10] The probability weight function is defined as

$$\pi(p_k) = \begin{cases} \dfrac{(p_k)^\gamma}{\left((p_k)^\gamma + (1-p_k)^\gamma\right)^{1/\gamma}} & x_k \geq 0 \\ \dfrac{(p_k)^\delta}{\left((p_k)^\delta + (1-p_k)^\delta\right)^{1/\delta}} & x_k < 0 \end{cases} \tag{5}$$

where γ and δ are the risk gain and loss attitude coefficients respectively.

Tversky and Kahneman [10] found that when $\alpha = \beta = 0.88, \theta = 2.25, \gamma = 0.61,$ $\delta = 0.72$, the experimental results are more consistent with the empirical results. To simplify calculation, we also take the above values in the paper.

2.2 Problem Description

The MSRDM process involves multiple stages and multiple states, which is usually risk, uncertainty and dynamic. Considering multiple reference points is helpful to make decisions results from multiple perspectives. This paper aims to select the desirable alternative(s) from a set of feasible alternatives according to the MSRDM problems.

In the MSRDM problem, let $A = \{a_i | i = 1, 2, \cdots, I\}$ be the set of I alternatives. Let $C = \{c_j | j = 1, 2, \ldots, J\}$ be the set of J attributes. Let $M = \{m^t | t = 1, 2, \ldots, T\}$ be the set of T stages. Let $S = \{s^n | n = 1, 2, \ldots, N\}$ be the set of N natural states. Let $WC = \{wc_1, wc_2, \ldots, wc_J\}$ be the weighting set of J attributes $\left(wc_j \in [0, 1], \sum_{j=1}^{J} wc_j = 1 \right),$ $WM = \{wm^1, wm^2, \ldots, wm^T\}$ be the weighting set of M stages $\left(wm^t \in [0, 1], \sum_{t=1}^{T} wm^t = 1 \right), P = \{p(s^1), p(s^2), \ldots, p(s^N)\}$ be the probability set of N states $\left(p(s^n) \in [0, 1], \sum_{n=1}^{N} p(s^n) = 1 \right)$. Let $X^m = (x_{ij}^{tn})_{I \times J}$ be the DM matrix in the stage m^t in the state s^n, which is showed in Table 1. $x_{ij}^{tn} = \left(Ex_{ij}^{tn}, En_{ij}^{tn}, He_{ij}^{tn} \right)$ is the decision value of alternative a_i with respect to attribute c_j in stage m^t in state s^n.

Table 1. MSRDM evaluation information.

| | | c_1 | | | | ... | c_J | | | |
		s^1	s^2	...	s^N	...	s^1	s^2	...	s^N
m^1	a_1	x_{11}^{11}	x_{11}^{12}	...	x_{11}^{1N}	...	x_{1J}^{11}	x_{1J}^{12}	...	x_{1J}^{1N}
	a_2	x_{21}^{11}	x_{21}^{12}	...	x_{21}^{1N}	...	x_{2J}^{11}	x_{2J}^{12}	...	x_{2J}^{1N}
	...	...	...	...	...	...	...	...	...	...
	a_I	x_{I1}^{11}	x_{I1}^{12}	...	x_{I1}^{1N}	...	x_{IJ}^{11}	x_{IJ}^{12}	...	x_{IJ}^{1N}
...	...	...	...	...	...	...	...	...	...	...
m^T	a_1	x_{11}^{T1}	x_{11}^{T2}	...	x_{11}^{TN}	...	x_{1J}^{T1}	x_{1J}^{T2}	...	x_{1J}^{TN}
	a_2	x_{21}^{T1}	x_{21}^{T2}	...	x_{21}^{TN}	...	x_{2J}^{T1}	x_{2J}^{T2}	...	x_{2J}^{TN}
	...	...	...	...	...	...	...	...	...	...
	a_I	x_{I1}^{T1}	x_{I1}^{T2}	...	x_{I1}^{TN}	...	x_{IJ}^{T1}	x_{IJ}^{T2}	...	x_{IJ}^{TN}

3 A MSRDM Method for Normal Cloud Model with Three Reference Points

3.1 The Setting Method of Three Reference Points

In MSRDM problems, the performance of an alternative will change dynamically as time goes on. Therefore, current situation, development trend and decision goal fluctuate with the change of stage. Considering a single reference point is difficult to evaluate the current situation, dynamic and inspiring nature comprehensively. Thus, it is difficult to systematically describe the development trend and characteristics.

The idea of setting the three reference points is showed in Fig. 1. The developmental reference point (DRP) is set by the performances of the previous stage. Compared with the DRP, the progress from the previous stage to the present stage can be obtained. The state reference point (SRP) is set by the expected performance of multiple states in the present stage. Unlike with the SRP, the extent to which one alternative is better than the expected performance of multiple states can be obtained. The promoting reference point (PRP) is set by the potential and the performance of the previous stage, which can be seen as the goal. The PRP can be used to adjust the degree and direction of the effort. Compared with the PRP, the degree of effort in the present stage can be obtained. In order to fully compare the advantages and disadvantages of MSRDM information, this paper sets up three reference points, i.e., development, multistate and promotion.

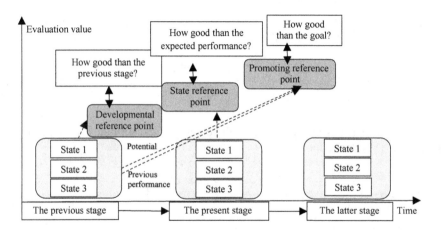

Fig. 1. The idea of setting the three reference points

State Reference Point 1. In multiple natural states, the expected value is the probability-weighted average of all possible values. It represents the central tendency of the values in multiple states. The expected value is what one expects to happen on average. If the value of an alternative is higher than the expected value, it will be a gain for the alternative to the expected value. The SRP is set by the average value of the expected values of alternatives in one stage.

Definition 6. The SRP of the attribute c_j at stage m^t is defined as

$$r_j^{ts} = \frac{1}{I}\sum_{i=1}^{I}\sum_{n=1}^{N}p(s^n)\cdot x_{ij}^{tn} = \left(Ex_j^{ts}, En_j^{ts}, He_j^{ts}\right) \tag{6}$$

Developmental Reference Point 2. From the viewpoint of development, the actual development level of an alternative at the present stage can be obtained by comparing the data with the data of the previous stage. The DRP is set by the performance of the previous stage. Thus, the progress of the present stage can be obtained by comparing it with the DRP. Compared with the previous stage, the greater the gain at the present stage, the better the development level of the present stage.

Definition 7. The DRP of the attribute c_j at stage m^t in natural state s^n is defined as

$$r_j^{tnd} = \frac{1}{I}\sum_{i=1}^{I} x_{ij}^{t-1,n} = \left(Ex_j^{tnd}, En_j^{tnd}, He_j^{tnd}\right) \tag{7}$$

Promoting Reference Point 3. From the viewpoint of promotion, reasonable goals should be set up to motivate people's subjective initiative. The PRP is set by the potential and performance of the previous stage. It actually is an estimate of the present stage based on the resources and historical foundations. The degree of realization of the PRPs reflects the degree of effort and the potential. Compared with the PRP, the greater the gain of the alternative, the better.

Definition 8. The maximum growth potential of the alternative a_i with respect to the attribute c_j at stage m^{t-1} in natural state s^n is defined as

$$\tau_{ij}^{t-1,n} = \frac{max_j x_{ij}^{t-1,n}}{x_{ij}^{t-1,n}} \tag{8}$$

Definition 9. The average maximum growth potential of the attribute c_j at stage m^{t-1} in natural state s^n is defined as

$$\tau_j^{t-1,n} = \frac{1}{I}\sum_{i=1}^{I} \left(\frac{max_j x_{ij}^{t-1,n}}{x_{ij}^{t-1,n}}\right) \tag{9}$$

Definition 10. The PRP of the attribute c_j at stage m^t in natural state s^n is defined as

$$r_j^{tnp} = \tau_j^{t-1,n}\frac{1}{I}\sum_{i=1}^{I} x_{ij}^{t-1,n} = \left(Ex_j^{tnp}, En_j^{tnp}, He_j^{tnp}\right) \tag{10}$$

3.2   The Value Function for Normal Cloud Model Under Three Reference Points

Comparing with three reference points can measure the performance of alternatives from different perspectives. The performance of one alternative at the current stage can be measured by comparing with the SRP. The performance of one alternative at the current stage can be measured by comparing with the SRP. The development performance of one alternative from the previous stage to the present stage can be measured by comparing with the DRP. Whether one alternative reaches the expected potential level can be measured by comparing with the GRP. When compared with the three reference points, the better the gain is, the better the alternative is. Then a value function is defined to obtain the gain or loss from a reference point. Take the DRP as an example.

Definition 11. The value function with the DRP is defined as

$$
v_{ij}^{tnd} = \begin{cases} \left(1 - sim\left(x_{ij}^{tn}, r_j^{tnd}\right)\right)^{\alpha} & Ex_{ij}^{tn} \geq Ex_j^{tnd} \\ -\theta\left(1 - sim\left(x_{ij}^{tn}, r_j^{tnd}\right)\right)^{\beta} & Ex_{ij}^{tn} < Ex_j^{tnd} \end{cases} \tag{11}
$$

where $sim\left(x_{ij}^{tn}, r_j^{tnd}\right)$ is the similarity between the attribute value and the DRP, which can be calculated by (2). When $Ex_{ij}^{tn} \geq Ex_j^{tnd}$, the value is a gain. The bigger the similarity between the attribute value and the DRP, the smaller the value of x_{ij}^{tn} as compared with the DRP r_j^{tnp}. When $Ex_{ij}^{tn} < Ex_j^{tnd}$, the value is a loss. The bigger the similarity between the attribute value and the DRP, the bigger the value of x_{ij}^{tn} compared with the DRP r_j^{tnp}.

The value functions for the SRP and the PRP are the same as the DRP.

3.3   The Aggregation Method for Three Kinds of Prospect Values

The prospect values with respect to each reference point can be calculated by Eq. (3). Then we can get $v_{ij}^{ts}, v_{ij}^{td}, v_{ij}^{tp}$. The comprehensive prospect values based on multiple reference points can be obtained by:

$$
v_{ij}^t = \lambda_j^{ts} \cdot v_{ij}^{ts} + \lambda_j^{td} \cdot v_{ij}^{td} + \lambda_j^{tp} \cdot v_{ij}^{tp} \tag{12}
$$

Where $\lambda_j^{ts}, \lambda_j^{td}, \lambda_j^{tp}\left(\lambda_j^{ts} + \lambda_j^{td} + \lambda_j^{tp} = 1, \lambda_j^{ts}, \lambda_j^{td}, \lambda_j^{tp} \in [0, 1]\right)$ represent the preference coefficients of different reference points at stage m^t. The preference coefficients can be given by decision makers in accordance with the actual situation. The coefficient of preference can also be determined according to the connections between the three kinds of prospect values.

The synthetic degree of grey incidence can describe the overall relationship of closeness between sequences [25]. So, we take the synthetic degree of grey incidence to obtain the preference coefficients of different reference points.

Let $X_j^{ts} = \left(v_{1j}^{ts}, v_{2j}^{ts}, \dots, v_{lj}^{ts}\right), X_j^{td} = \left(v_{1j}^{td}, v_{2j}^{td}, \dots, v_{lj}^{td}\right), X_j^{tp} = \left(v_{1j}^{tp}, v_{2j}^{tp}, \dots, v_{lj}^{tp}\right)$ be the sequences of the attribute c_j at stage m^t with respect to the three reference points.

The synthetic degree of grey incidence between each of the two-reference points can be calculated as $\rho_j^{tsd}, \rho_j^{tsp}, \rho_j^{tdp}$ [25]. Thus, the preference coefficients of different reference points can be obtained by:

$$
\lambda_j^{ts} = \frac{1}{2} \cdot \frac{\rho_j^{tsd} + \rho_j^{tsp}}{\rho_j^{tsd} + \rho_j^{tsp} + \rho_j^{tdp}} \tag{13}
$$

$$\lambda_j^{td} = \frac{1}{2} \cdot \frac{\rho_j^{tsd} + \rho_j^{tdp}}{\rho_j^{tsd} + \rho_j^{tsp} + \rho_j^{tdp}} \tag{14}$$

$$\lambda_j^{tp} = \frac{1}{2} \cdot \frac{\rho_j^{tsp} + \rho_j^{tdp}}{\rho_j^{tsd} + \rho_j^{tsp} + \rho_j^{tdp}} \tag{15}$$

The prospect values of alternatives at stage m^t can be obtained by

$$v_i^t = \sum_{j=1}^{J} wc_j \cdot v_{ij}^t \tag{16}$$

The prospect values of alternatives at all stages can be obtained by

$$v_i = \sum_{t=1}^{T} \sum_{j=1}^{J} wm^t \cdot wc_j \cdot v_{ij}^t \tag{17}$$

3.4   The Two-Stage Optimization Model

In MSRDM problems, the weights of attributes and stages can be given by decision makers. In some situations, it is difficult to determine the exact weights for decision-makers. Inappropriate weight setting may lead to errors in DM results. In this paper, we add the decision maker's judgment of weights to the priori information set. In this way, the weight optimization model can be more objective to determine the attribute weight, and we also take into account the influence of the subjective weights by DMs.

According to the idea of minimax reference point optimization [26], we designed the two-stage optimization model to solve the weights. The idea of the two-stage optimization model is showed in Fig. 2. The abscissa represents the weights of attributes, and the ordinate represents the weights of stages.

In the first stage of the two-stage optimization model, the maximum value of each alternative at each stage is obtained. In the second stage of the two-stage optimization model, the biggest distance between the maximum value and the actual value is minimized step by step. For example, the biggest distance in Fig. 2 is ε_3 at first. After optimization by modeling, the maximum distance is changed to ε_2. Finally, the biggest distance between the maximum value and the actual value is minimized. All the values of alternatives are as close as possible to the maximum value.

For the convenience of calculation, the prospect values of alternatives are standardized, the annotation is unchanged. Then we have $v_{ij}^t \in [0, 1]$.

The first stage model $M1$ is used to calculate the maximum value of each alternative at each stage. M1 is defined as

$$max(v_i^t) = max \sum_{j=1}^{J} wc_j \cdot v_{ij}^t \tag{18}$$

$$\begin{cases} \sum_{j=1}^{J} wc_j \cdot v_{ij}^t \leq 1, \ i = 1, 2, \ldots, I \\ wc_j \in H_1, \ j = 1, 2, \ldots, J \end{cases} \tag{19}$$

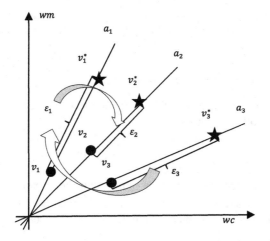

Fig. 2. The idea of the two-stage optimization method

The maximum value of each alternative at each stage can be calculated as v_i^{t*}. Each alternative can get its maximum value if model $M1$ is bounded. Because $v_{ij}^t \in [0, 1]$ and $wc_j \in [0, 1]$, the value of $\sum_{j=1}^{J} wc_j \cdot v_{ij}^t$ must be in the range $[0, 1]$. Attribute weights meet the prior set H_1, which can be expressed in 5 forms [27]. The prior set about attribute weights is usually decided by multiple decision makers.

The second stage model $M2$ aims to minimize the biggest distance between the maximum value and the actual value. M2 is defined as:

$$min(\varepsilon) \tag{20}$$

$$\begin{cases} \sum_{t=1}^{T} wm^t \cdot \left(v_i^{t*} - \sum_{j=1}^{J} wc_j \cdot v_{ij}^t \right) \le \varepsilon, i = 1, 2, \ldots, I \\ \sum_{t=1}^{T} \sum_{j=1}^{J} wm^t \cdot wc_j \cdot v_{ij}^t \le 1 \\ wc_j \in H_1, j = 1, 2, \ldots, J \\ wm^t \in H_2, t = 1, 2, \ldots, T \end{cases} \tag{21}$$

The minimum value of ε can be obtained from solving model $M2$. Constraints are the following conditions. The biggest distance between the maximum value and the actual value is less than or equal to ε. Each alternative is effective. Because $v_{ij}^t \in [0, 1]$, $wc_j \in [0, 1]$ and $wm^t \in [0, 1]$, the value of $\sum_{t=1}^{T} \sum_{j=1}^{J} wm^t \cdot wc_j \cdot v_{ij}^t$ must in the range $[0, 1]$. The attribute weight and the stage weight satisfy the prior set H_1 and H_2 respectively. The prior sets about attribute and stage weights are usually decided by multiple decision makers.

Then the final weight can be calculated as wm^{t*} and wc_j^*, the final ranking value can be calculated as v_i. The bigger the value of v_i, the better the alternative a_i is.

$$v_i = \sum_{t=1}^{T} \sum_{j=1}^{J} wm^{t*} \cdot wc_j^* \cdot v_{ij}^t \tag{22}$$

3.5 The DM Procedure of the MSRDM Method

The DM procedure to solve the MSRDM problems with three reference points is demonstrated in the following steps.

Step 1. Set three reference points.

The three reference points are obtained from (6)–(10).

Step 2. Calculate the prospect values.

Use (2) to calculate the similarity between x_{ij}^{tn} and the reference points. Then the prospect values under each reference point can be obtained from (11).

Step 3. Calculate the preference coefficients and the comprehensive prospect values.

Calculate the synthetic degree of grey incidence between each of the two-reference points [25]. Then, use (13)–(15) to calculate the preference coefficients, and use Eq. (12) to calculate the comprehensive prospect values.

Step 4. Solve the attribute weights and the stage weights.

Build model $M1$ (18)–(19) to get the maximum values v_i^{t*} of each alternative at different stages. Build model $M2$ (20)–(21) to solve the attribute weights and the stage weights.

Step 5. Calculate the final ranking value.

The final ranking values can be calculated from (22). The best alternative is $\max (v_i)$.

4 Numeral Example Analysis

4.1 Numeral Example Background

A pharmaceutical company carried out a risk assessment of the quality of products. There are 15 products to be evaluated, which comprise the set of alternatives $A = \{a_1, a_2, \ldots, a_I\}(I = 15)$. Evaluation attributes comprise the set of attributes $C = \{c_1, c_2, \ldots, c_J\}(J = 4)$. Three natural states comprise the set of natural states $S = \{s^n | n = 1, 2, \ldots, N\}(N = 3)$. The evaluation information from $T = 4$ stages comprises the decision-making matrix $\left(x_{ij}^{tn}\right)_{I \times J}$. Attribute weights meet the prior set $H_1 = \left\{\sum_{j=1}^{J} wc_j = 1; wc_j \geq 0.15; wc_1 + wc_3 \leq 0.45\right\}$, Stage weights meet the prior set $H_2 = \{\sum_{t=1}^{T} wm^t = 1; 0 < wm^1 \leq 0.2; wm^2 > 0.15; wm^1 + wm^2 \leq 0.45; wm^3 \geq 0.2; wm^4 \geq 0.3\}$.

There are four attributes to describe the products to be evaluated. c_1 represents the management level of raw material. It can be evaluated by decision makers with the linguistic sets (very good, good, medium, poor, very poor). c_2 represents the qualified rate of product quality. It can be obtained according to the data of product inspection. c_3 represents technological level. It can be evaluated by decision makers with the linguistic sets (very good, good, medium, poor, very poor). c_4 represents economic benefit. It can be obtained from the annual profit ratio.

There are three natural states: $s^1 = $ *the low* risks tate, $s^2 = $ *the medium* risk state, $s^3 = $ *the high* risk state. According to historical data, we have $P = \{p(s^1) = 0.65, p(s^2) = 0.25, p(s^3) = 0.1\}$. The DM matrixes at different stages are showed in the appendix as Tables 2, 3 and 4.

4.2   The Calculation Process

Step 1. Set three reference points.

(1) Use (6) to calculate the SRP in each stage.

Take the attribute c_1 as an example; the SRPs of attribute c_1 in four stages are showed in Fig. 3. The SRPs in stage m^1, m^2, m^3, m^4 are represented as $r_1^{1s} = (6.283, 3.431, 0.691), r_1^{2s} = (6.048, 3.32, 0.724), r_1^{1s} = (5.721, 3.214, 0.767), r_1^{1s} = (6.016, 3.342, 0.73)$.The symbols (".", "+", "Δ", "O") in Fig. 3 represent the SRP in stage m^1, m^2, m^3, m^4. Comparing the values of Ex in SRPs $(r_1^{1s}, r_1^{2s}, r_1^{3s}, r_1^{4s})$, we obtain that $Ex_1^{1s} > Ex_1^{2s} > Ex_1^{4s} > Ex_1^{3s}$. The SRP in stage m^1 is bigger than the other three reference points. The SRP in stage m^3 is smaller than the other three reference points.

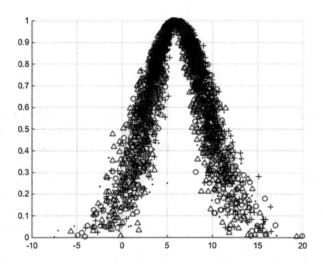

Fig. 3. The SRPs in different stages

(2) Use (7) to calculate the DRP in each state in each stage.

Take the attribute c_2 in stage m^1 as an example. The DRP in different states are represented   as $r_2^{11d} = (0.923, 0.069, 0.018), r_2^{12d} = (0.923, 0.042, 0.21), r_2^{13d} = (0.899, 0.065, 0.019)$. We can find that $Ex_2^{11d} \cong Ex_2^{12d} > Ex_2^{13d}$. But the cloud drops of r_2^{12d} are more dispersed than the other two $(He_2^{12d} > He_2^{13d} > He_2^{11d})$.

(3) Calculate the PRP in each state in each stage by Eq. (10) (omitted).

Take the attribute c_3 in stage m^1 as an example. The PRP in different states are represented as $r_3^{11p} = (8.253, 4.843, 0.926), r_2^{12p} = (8.404, 5.1, 0.948), r_2^{13p} = (6.839, 4.163, 0.949)$. We can find that $Ex_3^{12p} > Ex_3^{11p} > Ex_3^{13p}$.

Step 2. Calculate the prospect values.

Calculate the similarities and the prospect values (omitted). Take the prospect values in stage m^1 as an example. The prospect values are showed in Fig. 4. "s, d, p" in Fig. 4 represent the state, development and PRP.

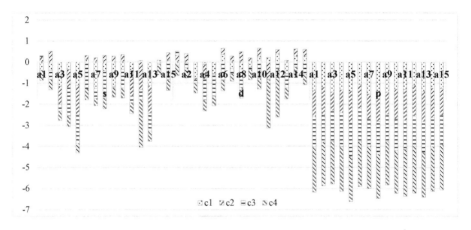

Fig. 4. The stacked column diagram of the prospect values in stage m^1

Compared with the SRPs, the prospect values of several alternatives with respect to attributes c_1 and c_2 are greater than 0, which means gains. This shows that the performances of these alternatives are higher than the expected performance of multiple states.

Compared with the DRPs, the prospect values of several alternatives with respect to attributes c_1, c_2 and c_3 are greater than 0, which means gains. This shows that the performances of these alternatives are higher than the levels of the previous stage. This means that these alternatives are working harder at the present stage than in the previous stage.

Compared with the PRPs, the prospect values are always lower than 0, which means losses. This shows that the performances of alternatives do not reach their potential. This means that these alternatives have not fully exploited their potential.

Step 3. Calculate the preference coefficients and the comprehensive prospect values.

Calculate the preference coefficients by using (13)–(15). The comprehensive prospect values based on multiple reference points can be obtained from (12) based on the preference coefficients (omitted).

Step 4. Calculate the attribute weights and the stage weights.

Solve model $M1$ and $M2$. The attribute weights and the stage weights can be obtained as $WC = \{0.15, 0.345, 0.15, 0.355\}$ and $WM = \{0.2, 0.25, 0.2, 0.35\}$. We can find that the weights of c_2 and c_4 are bigger than c_1 and c_3. The weights of stage m^4 is bigger than the other three stages.

Step 5. Calculate the final ranking value.

The final ranking value can be calculated from (22) and we get the ranking values of alternatives (as showed in Fig. 5).

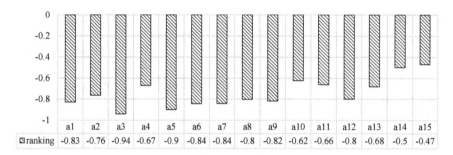

	a1	a2	a3	a4	a5	a6	a7	a8	a9	a10	a11	a12	a13	a14	a15
▨ranking	-0.83	-0.76	-0.94	-0.67	-0.9	-0.84	-0.84	-0.8	-0.82	-0.62	-0.66	-0.8	-0.68	-0.5	-0.47

Fig. 5. The ranking values of alternatives

Figure 5 shows that the best alternative is $a15$; the performances of alternative $a15$ are good under different reference points. The prospect values of alternative $a15$ are gains from the SRP and the DRP. This shows that the performance of alternative $a15$ is better than the expected performance of all alternatives in one stage and it is better than in the previous stages. The contribution of alternative $a15$ is greater, thus it is the best one.

5   Conclusions

The paper aims to propose a new MSRDM method for normal cloud model considering three reference points. The progress, current performance and the degree of effort are the criteria for measuring the multistage development of things. In this paper, the SRP is proposed to measure the current performance of an alternative under multiple states. The DRP is proposed to measure the progress in the previous stage. The PRP is proposed to measure the degree of effort and potential. Thus, a value function is defined to obtain the gain or loss from a reference point for normal cloud model. The prospect values under three reference points are aggregated by the synthetic degree of grey incidence. Then a two-stage weight optimization model is built to obtain the attribute weights and the stage weights based on the idea of minimax reference point optimization. The numeral example analysis shows its feasibility and validity for solving the MSRDM problems.

There are many interesting issues related to the problem with multiple reference points. The MCDM/GDM problems with multiple reference points will be researched in future. And a cloud model is a useful tool to deal with large quantity of data. We will develop a more appropriate method to deal with the cloud model.

Acknowledgements. This work was supported by the National Natural Science Foundation of China [71171112, 71502073, 71601002]; Funding of Jiangsu Innovation Program for Graduate Education ("the Fundamental Research Funds for the Central Universities") [KYZZ16_0147].

Appendix

	p1	p2	p3	p1	p2	p3
	c1			c2		
a1	6.69,2.27,0.35	5.7,1.93,0.47	5,1.82,0.55)	0.945,0.022,0.009	0.94,0.03,0.008	0.91,0.03,0.009
a2	8.07,2.75,0.19	5.7,1.93,0.47	5,1.82,0.55)	0.95,0.023,0.009	0.95,0.022,0.01	0.935,0.035,0.01
a3	5.7,1.93,0.47	5,1.82,0.55)	4.3,1.93,0.47	0.925,0.037,0.008	0.92,0.052,0.012	0.905,0.052,0.013
a4	6.69,2.27,0.35	5.7,1.93,0.47	5,1.82,0.55)	0.91,0.03,0.01	0.91,0.06,0.01	0.88,0.065,0.011
a5	5.7,1.93,0.47	5.7,1.93,0.47	5,1.82,0.55)	0.925,0.037,0.009	0.92,0.022,0.011	0.89,0.03,0.012
a6	8.07,2.75,0.19	6.69,2.27,0.35	5.7,1.93,0.47	0.945,0.037,0.008	0.94,0.037,0.01	0.905,0.038,0.011
a7	5,1.82,0.55)	5.7,1.93,0.47	5.7,1.93,0.47	0.94,0.015,0.009	0.94,0.03,0.01	0.91,0.033,0.011
a8	6.69,2.27,0.35	5,1.82,0.55)	5,1.82,0.55)	0.97,0.021,0.009	0.96,0.022,0.01	0.92,0.031,0.01
a9	5.7,1.93,0.47	5.7,1.93,0.47	5.7,1.93,0.47	0.94,0.024,0.008	0.94,0.037,0.009	0.915,0.041,0.01
a10	6.69,2.27,0.35	5.7,1.93,0.47	5.7,1.93,0.47	0.955,0.175,0.009	0.95,0.022,0.009	0.911,0.033,0.011
a11	6.69,2.27,0.35	5.7,1.93,0.47	5.7,1.93,0.47	0.9150.033,0.010	0.91,0.025,0.01	0.89,0.031,0.011
a12	8.07,2.75,0.19	6.69,2.27,0.35	5.7,1.93,0.47	0.925,0.028,0.009	0.92,0.023,0.009	0.905,0.025,0.011
a13	5.7,1.93,0.47	5.7,1.93,0.47	5,1.82,0.55)	0.93,0.035,0.01	0.92,0.025,0.01	0.905,0.031,0.01
a14	6.69,2.27,0.35	6.69,2.27,0.35	5.7,1.93,0.47	0.94,0.033,0.009	0.93,0.023,0.01	0.911,0.025,0.011
a15	6.69,2.27,0.35	6.69,2.27,0.35	5.7,1.93,0.47	0.955,0.031,0.009	0.94,0.028,0.009	0.91,0.031,0.011
	c3			c4		
a1	6.69,2.27,0.35	5.7,1.93,0.47	6.69,2.27,0.35	87.5,0.425,0.849	72.5,3.822,0.833	57.5,3.822,0.283
a2	8.07,2.75,0.19	5.7,1.93,0.47	5.7,1.93,0.47	85.5,0.425,0.849	74.5,1.274,0.849	65.5,0.425,1.416
a3	5.7,1.93,0.47	6.69,2.27,0.35	6.69,2.27,0.35	85,0.849,0.708	73.5,1.274,0.849	68,2.548,0.708
a4	5,1.82,0.55)	4.3,1.93,0.47	5,1.82,0.55)	87.5,1.274,0.566	72,2.548,0.425	65.4.247,0.142
a5	5,1.82,0.55)	4.3,1.93,0.47	5,1.82,0.55)	87,1.699,0.425	72,3.397,0.142	65.5,4.671,1.167
a6	5.7,1.93,0.47	5,1.82,0.55)	4.3,1.93,0.47	84.5,2.123,0.283	72.5,2.123,0.566	68.5,2.973,0.566
a7	8.07,2.75,0.19	5.7,1.93,0.47	5.7,1.93,0.47	82.5,2.123,0.283	76,0.849,0.991	71,0.849,1.274
a8	5.7,1.93,0.47	5,1.82,0.55)	5,1.82,0.55)	85,1.699,0.425	71.5,1.274,0.849	68,1.699,0.991
a9	6.69,2.27,0.35	8.07,2.75,0.19	5.7,1.93,0.47	84.5,2.973,0.667	73,1.699,0.708	67.5,2.973,0.566
a10	5,1.82,0.55)	6.69,2.27,0.35	5,1.82,0.55)	87,1.699,0.425	74.5,1.274,0.849	67,3.397,0.425
a11	5.7,1.93,0.47	8.07,2.75,0.19	6.69,2.27,0.35	80,0.849,0.708	69,2.548,0.425	63.5,3.822,0.283
a12	6.69,2.27,0.35	5.7,1.93,0.47	5.7,1.93,0.47	75,0.849,0.425	70,2.548,0.849	61.5,3.822,0.425
a13	5.7,1.93,0.47	6.69,2.27,0.35	4.3,1.93,0.47	85.5,2.123,0.283	72,0.849,0.991	61,1.699,0.991
a14	6.69,2.27,0.35	6.69,2.27,0.35	4.3,1.93,0.47	84,1.699,0.425	72.5,2.123,0.566	67.5,2.123,0.849
a15	5.7,1.93,0.47	8.07,2.75,0.19	5.7,1.93,0.47	84,1.699,0.425	72,2.548,0.47	64.5,4.671,1.167

Table 2. The evaluation information in stage m^2

with	p1	p2	p3	p1	p2	p3
	c1			c2		
a1	5.7,1.93,0.47	5.7,1.93,0.47	5.7,1.93,0.47	0.915,0.037,0.01	0.91,0.03,0.011	0.89,0.033,0.011
a2	6.69,2.27,0.35	5,1.82,0.55	5.7,1.93,0.47	0.91,0.045,0.01	0.93,0.045,0.009	0.91,0.045,0.011
a3	6.69,2.27,0.35	5,1.82,0.55	5.7,1.93,0.47	0.94,0.045,0.009	0.93,0.022,0.009	0.915,0.032,0.106
a4	6.69,2.27,0.35	5.7,1.93,0.47	5,1.82,0.55)	0.92,0.026,0.009	0.91,0.023,0.01	0.895,0.033,0.011
a5	6.69,2.27,0.35	5,1.82,0.55)	5,1.82,0.55)	0.905,0.038,0.009	0.9,0.022,0.01	0.88,0.025,0.011
a6	5.7,1.93,0.47	5.7,1.93,0.47	5,1.82,0.55)	0.91,0.035,0.01	0.9,0.037,0.012	0.85,0.045,0.014
a7	6.69,2.27,0.35	5.7,1.93,0.47	5,1.82,0.55	0.95,0.035,0.009	0.925,0.03,0.01	0.875,0.033,0.011
a8	5.7,1.93,0.47	5,1.82,0.55)	4.3,1.93,0.47	0.950.035,0.009	0.925,0.022,0.01	0.89,0.026,0.011
a9	5.7,1.93,0.47	5,1.82,0.55)	4.3,1.93,0.47	0.935,0.025,0.008	0.92,0.037,0.009	0.885,0.038,0.011
a10	6.69,2.27,0.35	5.7,1.93,0.47	5,1.82,0.55	0.95,0.015,0.009	0.93,0.022,0.01	0.89,0.035,0.011
a11	5.7,1.93,0.47	5.7,1.93,0.47	4.3,1.93,0.47	0.945,0.035,0.009	0.925,0.021,0.01	0.895,0.035,0.012
a12	6.69,2.27,0.35	6.69,2.27,0.35	5.7,1.93,0.47	0.93,0.025,0.009	0.925,0.023,0.01	0.887,0.035,0.011
a13	5.7,1.93,0.47	6.69,2.27,0.35	5,1.82,0.55	0.905,0.035,0.009	0.9,0.031,0.009	0.875,0.035,0.011
a14	8.07,2.75,0.19	5.7,1.93,0.47	5,1.82,0.55	0.955,0.021,0.008	0.94,0.031,0.009	0.915,0.035,0.01
a15	6.69,2.27,0.35	5,1.82,0.55	5.7,1.93,0.47	0.965,0.015,0.009	0.95,0.032,0.009	0.925,0.035,0.011
	c3			c4		
a1	5,1.82,0.55)	5.7,1.93,0.47	5,1.82,0.55)	83,0.849,1.416	72,1.699,0.566	66.5,3.822,0.566
a2	6.69,2.27,0.35	6.69,2.27,0.35	5.7,1.93,0.47	81,1.699,1.132	75,0.849,0.849	67,1.699,1.274
a3	5.7,1.93,0.47	6.69,2.27,0.35	5,1.82,0.55)	83,0.849,1.416	71,0.849,0.849	52.5,2.123,1.132
a4	6.69,2.27,0.35	8.07,2.75,0.19	5.7,1.93,0.47	83.5,2.123,0.991	72.5,1.274,0.708	67.5,2.123,1.132
a5	5.7,1.93,0.47	6.69,2.27,0.35	5,1.82,0.55)	86,1.699,1.132	74,0.849,0.849	65,3.397,0.708
a6	(5,1.82,0.55)	(5.7,1.93,0.47	(5.7,1.93,0.47	(77,0.849,1.416	(74,0.849,0.849	(62.5,2.973,0.849
a7	(5.7,1.93,0.47	(6.69,2.27,0.35	(4.3,1.93,0.47	(77,5.096,1.167	(76,2.548,0.283	(55.5,2.973,0.849
a8	(6.69,2.27,0.35	(6.69,2.27,0.35	(5,1.82,0.55)	(67.5,1.274,1.274	(59,1.699,0.566	(57,1.699,1.274
a9	(5.7,1.93,0.47	(6.69,2.27,0.35	(4.3,1.93,0.47	(82,0.849,1.416	(70,0.849,0.849	(60.5,2.973,0.849
a10	(6.69,2.27,0.35	(8.07,2.75,0.19	(5.7,1.93,0.47	(81.5,1.274,1.274	(70,3.397,,0.833	(59.5,5.521,1.167
a11	(5.7,1.93,0.47	(6.69,2.27,0.35	(4.3,1.93,0.47	(82,1.699,1.132	(71.5,2.123,0.425	(63.5,3.822,0.566
a12	(8.07,2.75,0.19	(8.07,2.75,0.19	(5,1.82,0.55)	(79.5,1.274,1.274	(73.5,2.123,0.425	(60.5,3.822,0.566
a13	(5.7,1.93,0.47	(6.69,2.27,0.35	(4.3,1.93,0.47	(80.5,2.973,0.708	(71,1.699,0.566	(61,1.699,1.274
a14	(8.07,2.75,0.19	(6.69,2.27,0.35	(5,1.82,0.55)	(81.5,3.822,0.425	(71,1.699,0.566	(63.5,4.671,0.283
a15	(6.69,2.27,0.35	(6.69,2.27,0.35	(5.7,1.93,0.47	(83.5,3.822,0.425	(70.5,1.274,0.708	(65,3.397,0.708

Table 3. The evaluation information in stage m^3

with	p1	p2	p3	p1	p2	p3
	c1			c2		
a1	5,1.82,0.55	5,1.82,0.55	4.3,1.93,0.47	0.975,0.015,0.009	0.96,0.027,0.01	0.935,0.031,0.011
a2	5.7,1.93,0.47	5,1.82,0.55	4.3,1.93,0.47	0.97,0.018,0.009	0.95,0.025,0.012	0.925,0.03,0.013
a3	5,1.82,0.55	5,1.82,0.55	4.3,1.93,0.47	0.91,0.055,0.008	0.91,0.037,0.01	0.89,0.04,0.012
a4	5.7,1.93,0.47	5.7,1.93,0.47	5,1.82,0.55)	0.97,0.014,0.01	0.96,0.035,0.011	0.93,0.04,0.011
a5	6.69,2.27,0.35	5,1.82,0.55	5.7,1.93,0.47	0.91,0.036,0.01	0.9,0.024,0.015	0.88,0.028,0.016
a6	5.7,1.93,0.47	5,1.82,0.55	5.7,1.93,0.47	0.915,0.023,0.008	0.91,0.035,0.012	0.875,0.04,0.013
a7	5.7,1.93,0.47	5.7,1.93,0.47	5,1.82,0.55)	0.925,0.016,0.008	0.92,0.045,0.01	0.089,0.046,0.011
a8	6.69,2.27,0.35	5.7,1.93,0.47	5,1.82,0.55	0.935,0.023,0.009	0.92,0.026,0.009	0.089,0.026,0.01

(*continued*)

Table 3. (*continued*)

with	p1	p2	p3	p1	p2	p3
a9	5.7,1.93,0.47	6.69,2.27,0.35	4.3,1.93,0.47	0.905,0.025,0.009	0.9,0.036,0.01	0.087,0.041,0.011
a10	6.69,2.27,0.35	5,1.82,0.55	4.3,1.93,0.47	0.935,0.035,0.009	0.92,0.042,0.011	0.885,0.045,0.012
a11	5,1.82,0.55	5,1.82,0.55	5.7,1.93,0.47	0.955,0.025,0.009	0.94,0.025,0.011	0.92,0.03,0.012
a12	6.69,2.27,0.35	5.7,1.93,0.47	5,1.82,0.55	0.925,0.031,0.009	0.93,0.032,0.014	0.915,0.035,0.015
a13	5.7,1.93,0.47	6.69,2.27,0.35	5.7,1.93,0.47	0.935,0.025,0.009	0.93,0.023,0.01	0.915,0.03,0.011
a14	6.69,2.27,0.35	6.69,2.27,0.35	5.7,1.93,0.47	0.943,0.011,0.007	0.94,0.023,0.01	0.92,0.025,0.011
a15	5.7,1.93,0.47	5.7,1.93,0.47	5,1.82,0.55	0.953,0.011,0.007	0.95,0.015,0.01	0.93,0.014,0.012
	c3			c4		
a1	5.7,1.93,0.47	6.69,2.27,0.35	5,1.82,0.55)	70.5,0.425,1.416	65,0.849,1.699	60.5,1.274,1.274
a2	6.69,2.27,0.35	6.69,2.27,0.35	5,1.82,0.55)	68.5,2.973,0.566	64.5,2.973,0.991	57.5,4.671,0.142
a3	5.7,1.93,0.47	6.69,2.27,0.35	5,1.82,0.55)	70,1.699,0.991	62,0.849,1.699	56,4.247,0.283
a4	5.7,1.93,0.47	5,1.82,0.55)	4.3,1.93,0.47	80,1.699,0.991	68.5,2.123,1.274	62.5,2.973,0.708
a5	5.7,1.93,0.47	5.7,1.93,0.47	4.3,1.93,0.47	77.5,2.123,0.849	66.5,2.973,0.991	59,5.096,1.333
a6	5,1.82,0.55)	5,1.82,0.55)	3.31,2.27,0.353	80.5,1.274,1.132	69.5,2.973,0.991	56.5,1.274,1.274
a7	5.7,1.93,0.47	5.7,1.93,0.47	5,1.82,0.55)	76,1.699,0.991	68.5,2.123,1.274	67.5,2.123,0.991
a8	6.69,2.27,0.35	6.69,2.27,0.35	5.7,1.93,0.47	76.5,2.973,0.566	70,4.247,0.566	62.5,2.123,0.991
a9	6.69,2.27,0.35	6.69,2.27,0.35	5.7,1.93,0.47	80,4.247,0.142	71.5,1.274,1.557	63,2.548,0.849
a10	8.07,2.75,0.19	6.69,2.27,0.35	5,1.82,0.55)	80,2.548,0.708	67.5,2.123,1.274	62,2.548,0.849
a11	5.7,1.93,0.47	6.69,2.27,0.35	4.3,1.93,0.47	77.5,2.123,0.849	71.5,1.274,1.557	67,1.699,1.132
a12	5.7,1.93,0.47	6.69,2.27,0.35	5,1.82,0.55)	78.5,2.973,0.566	73.5,1.274,1.557	66,2.548,0.849
a13	5,1.82,0.55)	5,1.82,0.55)	4.3,1.93,0.47	82.5,4.671,0.667	67.5,2.123,1.274	63,2.548,0.849
a14	5.7,1.93,0.47	5,1.82,0.55)	3.31,2.27,0.353	81.5,2.973,0.566	68,2.548,1.132	62.5,2.973,0.708
a15	6.69,2.27,0.35	6.69,2.27,0.35	4.3,1.93,0.47	82.5,2.123,0.849	70,5.945,1.333	62.5,3.822,0.425

Table 4. The evaluation information in stage m^4

with	p1	p2	p3	p1	p2	p3
	c1			c2		
a1	5.7,1.93,0.47	5,1.82,0.55	5,1.82,0.55	0.965,0.011,0.01	0.96,0.025,0.011	0.945,0.035,0.012
a2	5.7,1.93,0.47	5.7,1.93,0.47	5,1.82,0.55	0.965,0.012,0.01	0.955,0.026,0.013	0.915,0.03,0.021
a3	8.07,2.75,0.19	6.69,2.27,0.35	5.7,1.93,0.47	0.921,0.045,0.009	0.92,0.034,0.014	0.89,0.36,0.018
a4	6.69,2.27,0.35	5.7,1.93,0.47	5,1.82,0.55	0.95,0.012,0.009	0.945,0.021,0.014	0.925,0.03,0.015
a5	5.7,1.93,0.47	5.7,1.93,0.47	5,1.82,0.55	0.92,0.033,0.011	0.91,0.022,0.014	0.895,0.032,0.019
a6	6.69,2.27,0.35	8.07,2.75,0.19	6.69,2.27,0.35	0.91,0.025,0.011	0.91,0.042,0.012	0.855,0.045,0.015
a7	5,1.82,0.55	5.7,1.93,0.47	5,1.82,0.55	0.93,0.011,0.01	0.92,0.035,0.011	0.895,0.038,0.016
a8	6.69,2.27,0.35	5.7,1.93,0.47	5.7,1.93,0.47	0.94,0.025,0.01	0.935,0.025,0.015	0.89,0.035,0.018
a9	5,1.82,0.55	5.7,1.93,0.47	5,1.82,0.55	0.92,0.026,0.01	0.92,0.024,0.012	0.85,0.035,0.015
a10	5.7,1.93,0.47	5,1.82,0.55	5,1.82,0.55	0.94,0.025,0.01	0.93,0.023,0.011	0.89,0.055,0.021
a11	5,1.82,0.55	6.69,2.27,0.35	5.7,1.93,0.47	0.96,0.02,0.01	0.935,0.012,0.012	0.875,0.034,0.015
a12	6.69,2.27,0.35	5.7,1.93,0.47	5,1.82,0.55	0.93,0.025,0.012	0.925,0.015,0.014	0.885,0.025,0.018
a13	6.69,2.27,0.35	5.7,1.93,0.47	5,1.82,0.55	0.95,0.02,0.01	0.93,0.032,0.01	0.905,0.045,0.015
a14	5.7,1.93,0.47	6.69,2.27,0.35	5.7,1.93,0.47	0.95,0.015,0.012	0.925,0.012,0.01	0.89,0.025,0.018
a15	6.69,2.27,0.35	6.69,2.27,0.35	5.7,1.93,0.47	0.958,0.012,0.01	0.925,0.025,0.011	0.885,0.031,0.016
	c3			c4		
a1	5.7,1.93,0.47	6.69,2.27,0.35	4.3,1.93,0.47	71.5,2.973,1.132	64,4.247,0.283	60,4.247,0.142

(*continued*)

Table 4. (*continued*)

with	p1	p2	p3	p1	p2	p3
a2	5.7,1.93,0.47	6.69,2.27,0.35	5,1.82,0.55)	69.5,6.37,1.167	60,4.247,0.283	54,4.247,0.142
a3	6.69,2.27,0.35	6.69,2.27,0.35	5,1.82,0.55)	69,3.397,0.991	64,5.096,1.167	55.5,4.671,1.167
a4	8.07,2.75,0.19	8.07,2.75,0.19	5.7,1.93,0.47	77,1.699,1.557	70,1.699,1.132	62.5,2.123,0.849
a5	5.7,1.93,0.47	6.69,2.27,0.35	5,1.82,0.55)	79,2.548,1.274	69.5,1.274,1.274	62,2.548,0.708
a6	5,1.82,0.55)	6.69,2.27,0.35	5,1.82,0.55)	84.5,3.822,0.849	72.5,2.123,0.991	66.5,2.973,0.566
a7	5,1.82,0.55)	5,1.82,0.55)	4.3,1.93,0.47	76,0.849,1.84	73.5,2.973,0.708	67.5,2.123,0.849
a8	6.69,2.27,0.35	6.69,2.27,0.35	5,1.82,0.55)	72.5,2.123,1.416	69,4.247,0.283	61,2.548,0.708
a9	5,1.82,0.55)	5.7,1.93,0.47	5,1.82,0.55)	78,5.096,0.425	74,3.397,0.566	61.5,2.973,0.566
a10	5.7,1.93,0.47	5.7,1.93,0.47	4.3,1.93,0.47	78.5,2.973,1.132	70,1.699,1.132	60,4.247,0.142
a11	6.69,2.27,0.35	8.07,2.75,0.19	5.7,1.93,0.47	78,2.548,1.274	72.5,2.123,0.991	67,1.699,0.991
a12	6.69,2.27,0.35	6.69,2.27,0.35	5,1.82,0.55)	75.5,2.973,1.132	72.5,3.822,0.425	67.5,1.274,1.132
a13	8.07,2.75,0.19	8.07,2.75,0.19	5.7,1.93,0.47	81.5,2.973,1.132	72.5,2.123,0.991	61.5,2.973,0.566
a14	5.7,1.93,0.47	6.69,2.27,0.35	5,1.82,0.55)	82,1.699,1.557	74,3.397,0.566	67.5,2.123,0.849
a15	6.69,2.27,0.35	6.69,2.27,0.35	4.3,1.93,0.47	78.5,1.274,1.699	73,2.548,0.849	67.5,2.123,0.849

References

1. Pospichal, J., Kvasnicka, V.: Multistage decision-making using simulated annealing applied to a fuzzy automaton. Appl. Soft Comput. **2**(2), 140–151 (2003)
2. Sirbiladze, G., Khutsishvili, I., Ghvaberidze, B.: Multistage decision-making fuzzy methodology for optimal investments based on experts' evaluations. Eur. J. Oper. Res. **232**(1), 169–177 (2014)
3. Guo, P., Li, Y.: Approaches to multistage one-shot decision making. Eur. J. Oper. Res. **236**(2), 612–623 (2014)
4. Qu, J., Meng, X., You, H.: Multi-stage ranking of emergency technology alternatives for water source pollution accidents using a fuzzy group decision making tool. J. Hazard. Mater. **310**, 68–81 (2016)
5. Zhang, S.W., Guo, H.X., Zhu, K.J., Yu, S.W., Li, J.L.: Multistage assignment optimization for emergency rescue teams in the disaster chain. Knowl. Based Syst. **137**, 123–137 (2017)
6. Yoon, K.: A reconciliation among discrete compromise solutions. J. Oper. Res. Soc. **38**(3), 277–286 (1987)
7. Walczak, D., Rutkowska, A.: Project rankings for participatory budget based on the fuzzy TOPSIS method. Eur. J. Oper. Res. **260**(2), 706–714 (2017)
8. Tavana, M., Caprio, D.D., Santos-Arteaga, F.J.: An extended stochastic VIKOR model with decision maker's attitude towards risk. Inf. Sci. **432**, 301–318 (2018)
9. Kahneman, D., Tversky, A.: Prospect theory: an analysis of decision under risk. Econometrica **47**, 263–291 (1979)
10. Tversky, A., Kahneman, D.: Advances in prospect theory: cumulative representation of uncertainty. J. Risk Uncertain. **5**, 297–323 (1992)
11. Wang, L., Zhang, Z.X., Wang, Y.M.: A prospect theory-based interval dynamic reference point method for emergency DM. Expert Syst. Appl. **42**(23), 9379–9388 (2015)
12. Zhu, J., Ma, Z., Wang, H., Chen, Y.: Risk decision-making method using interval numbers and its application based on the prospect value with multiple reference points. Inf. Sci. **385–386**, 415–437 (2017)

13. Cables, E., Lamata, M.T., Verdegay, J.L.: RIM-reference Ideal method in multicriteria decision making. Inf. Sci. **337–338**, 1–10 (2016)
14. Mandal, P., Kaul, R., Jain, T.: Stocking and pricing decisions under endogenous demand and reference point effects. Eur. J. Oper. Res. **264**, 181–199 (2018)
15. Wang, J.Q., Peng, L., Zhang, H.Y., Chen, X.H.: Method of multi-criteria group decision-making based on cloud aggregation operators with linguistic information. Inf. Sci. **274**, 177–181 (2014)
16. Li, D., Han, J., Shi, X., Chan, M.C.: Knowledge representation and discovery based on linguistic atoms. Knowl. Based Syst. **10**(7), 431–440 (1998)
17. Yang, X., Zeng, L., Luo, F., Wang, S.: Cloud hierarchical analysis. J. Inf. Comput. Sci. **12**, 2468–2477 (2010)
18. Yang, X., Yan, L., Zeng, L.: How to handle uncertainties in AHP: the cloud delphi hierarchical analysis. Inf. Sci. **222**(3), 384–404 (2013)
19. Peng, B., Zhou, J., Peng, D.: Cloud model based approach to group decision making with uncertain pure linguistic information. J. Intell. Fuzzy Syst. **32**(3), 1959–1968 (2017)
20. Wang, D., Liu, D., et al.: A cloud model-based approach for water quality assessment. Environ. Res. **149**, 113–121 (2016)
21. Xu, C.L., Wang, G.Y.: A novel cognitive transformation algorithm based on gaussian cloud model and its application in image segmentation. Algorithms **76**(4), 1039–1070 (2017)
22. Zhang, R.L., Shan, M.Y., Liu, X.H., Zhang, L.H.: A novel fuzzy hybrid quantum artificial immune clustering algorithm based on cloud model. Eng. Appl. AI **35**, 1–13 (2014)
23. Li, D., Liu, C., Gan, W.: A new cognitive model: cloud model. Int. J. Intell. Syst. **24**(3), 357–375 (2009)
24. Wang, J., Zhu, J., Liu, X.: An integrated similarity measure method for normal cloud model based on shape and distance. Syst. Eng. Theory Pract. **37**(3), 742–751 (2017b). (in Chinese)
25. Liu, S., Lin, Y.: Grey Information: Theory and Practical Applications. Springer, London (2006). https://doi.org/10.1007/1-84628-342-6
26. Yang, J.B.: Minimax reference point approach and its application for multiobjective optimisation. Eur. J. Oper. Res. **126**(3), 90–105 (2000)
27. Kim, S.H., Choi, S.H., Kim, J.K.: An interactive procedure for multiple attribute group decision making with incomplete information: range-based approach. Eur. J. Oper. Res. **118**(1), 139–152 (1999)

System Portfolio Selection Under Hesitant Fuzzy Information

Zhexuan Zhou, Xiangqian Xu, Yajie Dou[(✉)], Yuejin Tan,
and Jiang Jiang

College of Systems Engineering, National University of Defense Technology,
Changsha 410073, Hunan, China
yajiedou_nudt@163.com

Abstract. System portfolio selection faces multi-criteria and multi-objective problems, which lead the decision-makers to build a decision model. Otherwise, the system evaluation value is not clear and the multi-objective of the system is difficult to outweigh. To solve the problem, a value-risk ratio model with Hesitant Fuzzy Set (HFS) is used for portfolio selection. To be specific, in this model, the HFS is used to evaluate the value and risk of systems; and the portfolio value and portfolio risk are calculated with HFS operation. Meanwhile, the value-risk rate is applied to address the problem of multi-objective for system portfolio. Finally, one numerical example for system portfolio selection is given to illustrate the applicability of the proposed model.

Keywords: System portfolio selection · Hesitant Fuzzy Set · Decision-making
Value and risk model

1 Introduction

Markowitz portfolio selection theory [1], has been widely used in project selection [2], including medical capital budgeting [3] and defense acquisition [4]. In the field of management science and operations research, portfolio decision theory as a resource optimization method was more widely used in R & D (Project R & D). Portfolio Decision Analysis helps decision makers choose from alternative options by analyzing the relevant constraints and preferences in the decision-making process.

A number of researchers applied the portfolio selection theory in military system selection. Zhou et al. [5] proposed a hybrid approach based on portfolio selection theory for weapon system selection combined with weapons manufacturing. Dou et al. [6] proposed a portfolio selection model with demand-pull to solve the multi-weapon systems problem in the defense manufacturing process. The Department of Defense (DoD) used portfolio selection methods to make the plan for defense manufacturing cost [7]. Zhou et al. [8] used the fuzzy cluster in system portfolio selection. However, the evaluation for the weapon system portfolio selection is difficult to describe. In many cases, experts evaluate the values with hesitation. In this study, the HFSs are used to describe evaluation of value and risk.

Zadeh [9] proposed the Fuzzy set in 1965. From that time, many researchers have conducted research in Fuzzy set in decision-making [10, 11]. Recently, Torra [12] has

© Springer International Publishing AG, part of Springer Nature 2018
Y. Chen et al. (Eds.): GDN 2018, LNBIP 315, pp. 33–40, 2018.
https://doi.org/10.1007/978-3-319-92874-6_3

proposed Hesitant Fuzzy Set (HFS) which plays an important part in decision-making area. Then, Xu and Xia [13] developed the math form of Hesitant Fuzzy Set and defined related operations. Farhadinia [14, 15] proposed a series of Scoring Function and ranking methods for hesitant fuzzy numbers. Zhao *et al.* [16] proposed a hesitant fuzzy multi-attribute decision-making approach with the minimum deviation. Zhu *et al.* [17] discussed the ranking methods with hesitant fuzzy preference relations in the group decision making. Zhang [18] built a framework for group decision making with hesitant fuzzy preference relations using the multiplicative consistency. In other words, HFS has been widely used in decision-making.

The evaluation of value and risk are vague in system portfolio selection, which makes it difficult for decision-makers to describe the evaluation data in the selection model. The traditional method used the real number to model, which cannot accurately reflect the value and risk of the system. HFS is a useful approach to describe the fuzzy information in group decision making. In this study, HFS was used for system portfolio selection, which better represents real-world decision-making processes. This was the major research motivation for this study.

The main contribution of this study is solving system portfolio selection by using the HFS. In the decision-making process, value and risk are mainly considered. HFS is used to represent value and risk evaluation for the weapon system. Then, the portfolio value and portfolio risk are calculated with HFS operation. To gain the consistent order of weapon system portfolios, the value–risk rate model is built. Finally, a numerical example of a system portfolio selection is used to illustrate the advantages of the method.

The remainder of the paper is structured as follows: Sect. 2 is the preliminaries for HFS. Section 3 is a detailed description of hesitant fuzzy portfolio selection model with value-rate ratio. In this model, the portfolio value and risk are calculated by using the HFS operation. Sections 4 is a case study for weapon system portfolio selection. Section 5 is a discussion and conclusions of this study.

2 Preliminaries

To describe the decision-making process clearly, generally, people use the hesitant fuzzy information to represent objects and concepts. Recently, Zhang and Xu [20] defined the Hesitant Fuzzy Set, which includes the membership and non-membership degrees of several different values, respectively. A hesitant fuzzy set A on X is a function H(A) that when applied to X returns a finite subset of [0, 1], which can be represented as the following mathematical symbol.

Definition 1 [13]. HFS can be described by

$$H = \{ <x, h_A(x) > | x \in X \} \tag{1}$$

where $H_A(x)$ is a set of some values in [0, 1], denoting the possible membership degrees of the element x to the set A. Xia and Xu [13] named $h_A(x)$ a Hesitant Fuzzy Element (HFE).

Definition 2 [12, 13]. According to the definition of hesitation fuzzy number, the following operation can be gained. Supposing that there are three hesitant fuzzy numbers h_1, h_2, h_3, we define the following operations:

$$h_1 \oplus h_2 = H\{(\gamma_1 + \gamma_2 - \gamma_1\gamma_2) \mid \gamma_1 \in h_1, \gamma_2 \in h_2\},$$
$$h_1 \otimes h_2 = H\{\gamma_1\gamma_2 \mid \gamma_1 \in h_1, \gamma_2 \in h_2\},$$
$$h_1 \cup h_2 = H\{\max(\gamma_1, \gamma_2) \mid \gamma_1 \in h_1, \gamma_2 \in h_2\},$$
$$h_1 \cap h_2 = H\{\min(\gamma_1, \gamma_2) \mid \gamma_1 \in h_1, \gamma_2 \in h_2\},$$
$$\lambda h = H\{1 - (1 - \gamma)^\lambda \mid \gamma \in h\}(\lambda > 0), \text{ and}$$
$$h^\lambda = H\{\gamma^\lambda \mid \gamma \in h\}(\lambda > 0).$$

Assumption [19]. To make the hesitant fuzzy number easy o calculate, the hesitant number which has fewer elements should be extended until all hesitant number elements are the same. If the decision maker is pursuing risk, then the largest element of hesitation should be added to all corresponding elements until the two hesitant fuzzy numbers are equal, and vice versa.

Definition 3 [20]. If there is a hesitant fuzzy number $h = H\{\gamma_1, \gamma_2, \ldots, \gamma_n\}$, then the hesitant fuzzy score function is

$$Z(h) = (((\gamma_1)^\delta + (\gamma_1)^\delta + \cdots + (\gamma_n)^\delta)/n)^{1/\delta}. \tag{2}$$

Here, this hesitant fuzzy score function is utilized to calculate the hesitant fuzzy number, because this function can compare hesitant fuzzy numbers clearly and the process of calculation is easy. On the other hand, users can adjust this parameter δ according to their own preferences during the real decision-making in this function.

3 Hesitant Fuzzy Portfolio Selection Model with V-R Ratio

The optimization model of a system portfolio selection is a commonly maximized value (technical maturity level and demand satisfaction level) or a minimized cost (cost, expense, resources and risk, etc.). This model takes the system and the portfolio as the value of the consideration objects. The criteria of these different dimensions are mapped into a certain model structure, and the problem is transformed into a single objective 0–1 programming problem or a multi-objective 0–1 programming problem. At present, the most common and effective practice is using the multi-criteria evaluation model and the additive value function to define the combination selection.

According to the portfolio rules, if there are n systems, then 2^{n-1} portfolios can be gained.

This study also follows the value model, taking value and risk as the main consideration in the process of weapon combination selection. Value and risk are evaluated by experts using HFS, which is closer to the real decision-making processing.

Definition 4. Let V_{11} be the portfolio value of systems. It can be described as

$$V_{ij} = v_i \oplus v_j \tag{3}$$

where V_{ij} is the portfolio value of systems, and v_i, v_j are the value of systems i, j.
If the value is described by hesitant fuzzy number, then

$$V_{ij} = h_i^v \oplus h_j^v = H\{\gamma_i^v + \gamma_j^v - \gamma_i^v \gamma_j^v, \gamma_i^v \in h_i^v, \gamma_j^v \in h_j^v\} \tag{4}$$

where h_i^v are the elements in value set V_{ij}.

Definition 5. Let R_{ij} be the portfolio risk of systems. It can be described as

$$R_{ij} = r_i \cup r_j \tag{5}$$

where R_{ij} is the portfolio risk of systems, and r_i, r_j are the risk of systems i, j.
If the risk is described by hesitant fuzzy number, then

$$R_{ij} = h_i^r \cup h_j^r = H\{\max(\gamma_i^r, \gamma_j^r), \gamma_i^r \in h_i^r, \gamma_j^r \in h_j^r\} \tag{6}$$

where r_i^v are the elements in risk set R_{ij}. Hence, a multi-object model can be built.

Model 1.

$$\begin{cases} Max \quad V(m, z(m, i)) = \sum_{i=1}^{m} v_j(x_i) \bullet z(m, i) \\ Min \quad R(m, z(m, i)) = \bigcup_{i=1}^{m} r_j(x_i) \bullet z(m, i) \end{cases} \tag{7}$$

where m is the number of weapon systems; $z(m, i)$ belongs to $\{0, 1\}$, if and only if $z(m, i) = 1$.

It shows that when the number of weapon systems is m, the i-th weapon system is selected for the portfolio. $v_j(x_i)$ is the system value and $r_j(x_i)$ is the risk of the system. Model 1 seeks to maximize the value of the system and minimize the risk. After analysis, the above model can be transformed into Model 2.

Model 2.

$$VR = \frac{V(m, z(m, i))}{R(m, z(m, i))} = \frac{\sum_{i=1}^{m} v_j(x_i) \bullet z(m, i)}{\bigcup_{i=1}^{m} r_j(x_i) \bullet z(m, i)} \tag{8}$$

VR shows the ratio of value to risk. Through this ratio of cost and risk, the weapon system portfolio can be ranked.

The process of hesitant fuzzy portfolio selection model with V-R ratio is as follows:

Step 1: The value and risk value of the hesitation and fuzzy evaluation of the expert system are obtained.

Step 2: The extension rules of Xu and Xia [19] are used to expand the fuzzy number of evaluation hesitation.

Step 3: During the weapon system portfolio selection, (4) and (6) are used to calculate the value and risk of the portfolio.

Step 4: Use (2) to calculate the score value Z of the hesitant fuzzy number of the combined value and the risk.

Step 5: Model 2 is used to calculate the ratio between value and risk. The system portfolio selection is ranked by the ratio.

Step 6: End.

4 Case Study

To illustrate the applicability of the proposed approach, the numerical example of weapon system portfolio selection is presented. Let us suppose that one decision-maker wants to make a decision on portfolio selection from three weapon systems (s_1, s_1, s_1).

To make the decision convenient and efficient, the value and the risk of the weapon system are mainly considered. Hence, there are 7 weapon system portfolios that can be gained. Then, the decision-maker ranks the 7 weapon system portfolios and selects the best one. The specific data are Table 1.

Table 1. Value and risk for systems

S (System)	S_1	S_2	S_3
V (value)	H{0.5, 0.6}	H{0.4, 0.5, 0.6}	H{0.6, .65}
Risk	H{0.4, 0.3}	H{0.5, 0.5}	H{0.55, 0.6, 0.7}

According to the extension rule given in *Assumption 1*, in this example, the risk is opposed, and the smaller elements of hesitation fuzzy numbers are expanded here (Table 2).

Table 2. Extended value and risk

S (System)	S_1	S_2	S_3
V (value)	H{0.5, 0.5, 0.6}	H{0.4, 0.5, 0.6}	H{0.6, 0.6, .65}
Risk	H{0.3, 0.3, 0.4}	H{0.5, 0.5, 0.5}	H{0.55, 0.6, 0.7}

According to the portfolio rules there can be seven portfolios P_1–P_7, Then accordingly (4)–(6) can be used to calculate the value and risk value of these seven portfolios, as showed in Table 3.

Table 3. System portfolio selection

System portfolio	P₁	P₂	P₃	P₄	P₅	P₆	P₇
System	S₁	S₂	S₃	S₁, S₂	S₁, S₃	S₂, S₃	S₁, S₂, S₃
V (value)	H{0.5, 0.5, 0.6}	H{0.4, 0.5, 0.6}	H{0.6, 0.6, .65}	H{0.7, 0.75, 0.84}	H{0.8, 0.8, 0.86}	H{0.76, 0.8, 0.86}	H{0.88, 0.9, 0.94}
Risk	H{0.4, 0.45, 0.55}	H{0.5, 0.5, 0.5}	H{0.55, 0.6, 0.7}	H{0.85, 0.86, 0.91}	H{0.69, 0.72, 0.82}	H{0.868, 0.89, 0.933}	H{0.92, 0.94, 0.97}

Score values for the hesitant fuzzy number of portfolio value and risk are calculated by formulating (2); and the results are shown in Table 4.

Table 4. Score values for the hesitant fuzzy number of portfolios

Portfolios	P₁	P₂	P₃	P₄	P₅	P₆	P₇
Z (Value)	0.5315	0.4939	0.6163	0.7614	0.8196	0.8057	0.9064
Z (Risk)	0.5134	0.5	0.6139	0.5315	0.6139	0.6139	0.6139

Hence, the value order and risk order of weapon system portfolios are clearly listed. They are described in Figs. 1 and 2 respectively.

Figures 1 and 2, show the order of portfolios value and risk clearly. However, the value and risk are two conflict objects. Thus, it is difficult to rank the weapon system portfolios. To gain the consistent order for weapon system portfolios, the ratio between value and risk is calculated by (8); and the result is shown in Table 5.

The consistent order is described in Fig. 3. Figure 3, shows that the weapon system portfolio P₇ is the best, then P₄ is second, and P₂ is the worst.

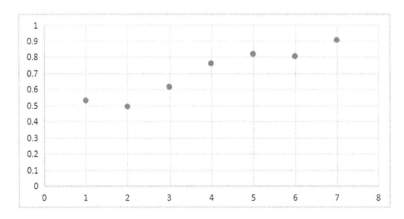

Fig. 1. Value order for weapon system portfolios

Table 5. Ratio between value and risk

Portfolios	P_1	P_2	P_3	P_4	P_5	P_6	P_7
Ratio	1.04	0.99	1.00	1.43	1.34	1.31	1.48

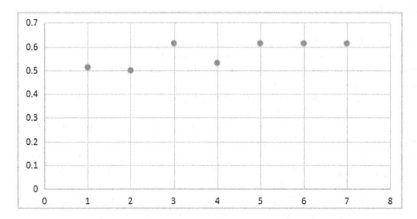

Fig. 2. Risk order for weapon system portfolios

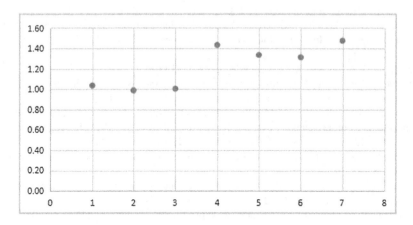

Fig. 3. The consistent order for weapon system portfolio

5 Conclusions

System portfolio selection is based on research. The system evaluation value is fuzzy and the system portfolio faces multiple objectives. To achieve the best portfolio, the value and risk are considered in this study and the HFS is used to evaluate the system. It should be mentioned that the HFS operation is used to calculate the portfolio value and risk. Moreover, value-risk ratio for the system is utilized to eliminate multi-object.

Acknowledgments. We are thankful to the Editor and the reviewers for their valuable comments and detailed suggestions to improve the presentation of the paper. Further, we also acknowledge the support in part by the National Natural Science Foundation of China under Grant No. 71690233, No. 71671186 and No. 71401167.

References

1. Markowitz, H.: Portfolio selection. J. Financ. **7**(1), 77–91 (1952)
2. Stummer, C., Heidenberger, K.: Interactive R&D portfolio analysis with project interdependencies and time profiles of multiple objectives. IEEE Trans. Eng. Manag. **50**, 175–183 (2003)
3. Peacock, J., Richardson, R., Carter, D.: Edwards, priority setting in health care using multi-attribute utility theory and programme budgeting and marginal analysis (PBMA). Soc. Sci. Med. **64**, 897–910 (2007)
4. Buede, D., Bresnick, T.: Applications of decision analysis to the military systems acquisition process. Interfaces **22**(6), 110–125 (1992)
5. Zhou, Y., Jiang, J., Yang, Z., Tan, Y.: A hybrid approach for multi-weapon production planning with large-dimensional multi-objective in defence manufacturing. Proc. Inst. Mech. Eng. Part B J. Eng. Manuf. **228**(2), 302–316 (2014)
6. Dou, Y., Zhang, P., Ge, B., Jiang, J., Chen, Y.: An integrated technology pushing and requirement pulling model for weapon system portfolio selection in defence acquisition and manufacturing. Proc. Inst. Mech. Eng. Part B J. Eng. Manuf. **229**(6), 1781–1789 (2014)
7. Katz, D.R., Sarkani, S., Mazzuchi, T., Conrow, E.H.: The relationship of technology and design maturity to DoD weapon system cost change and schedule change during engineering and manufacturing development. Syst. Eng. **18**(1), 1–15 (2015)
8. Zhou, Z., Dou, Y., Xia, B., et al.: Weapon systems portfolio selection based on fuzzy clustering analysis. In: IEEE International Conference on Control Science and Systems Engineering, pp. 702–705. IEEE (2017)
9. Zadeh, L.A.: Fuzzy sets. Inf. Control **8**(3), 338–353 (1965)
10. Xia, M., Xu, Z., Chen, N.: Some hesitant fuzzy aggregation operators with their application in group decision-making. Group Decis. Negot. **22**, 259–279 (2013)
11. Chen, N., Xu, Z.S., Xia, M.M.: Interval-valued hesitant preference relations and their applications to group decision making. Knowl.-Based Syst. **37**, 528–540 (2013)
12. Torra, V.: Hesitant fuzzy sets. Int. J. Intell. Syst. **25**(6), 529–539 (2010)
13. Xia, M., Xu, Z.: Hesitant fuzzy information aggregation in decision making. Int. J. Approx. Reason. **52**(3), 395–407 (2011)
14. Farhadinia, B.: A novel method of ranking hesitant fuzzy values for multiple attribute decision-making problems. Int. J. Intell. Syst. **28**(8), 752–767 (2013)
15. Farhadinia, B.: A series of score functions for hesitant fuzzy sets. Inf. Sci. **277**(2), 102–110 (2014)
16. Zhao, H., Xu, Z., Wang, H., Liu, S.: Hesitant fuzzy multi-attribute decision-making based on the minimum deviation method. Soft Comput. **21**(12), 3439–3459 (2016)
17. Zhu, B., Xu, Z., Xu, J.: Deriving a ranking from hesitant fuzzy preference relations under group decision making. IEEE Trans. Cybern. **44**(8), 1328–1337 (2017)
18. Zhang, Z.: A framework of group decision making with hesitant fuzzy preference relations based on multiplicative consistency. Int. J. Fuzzy Syst. **19**(4), 1–15 (2016)
19. Xu, Z., Xia, M.: Distance and similarity measures for hesitant fuzzy sets. Inf. Sci. **181**(11), 2128–2138 (2011)
20. Zhang, X., Xu, Z.: The TODIM analysis approach based on novel measured functions under hesitant fuzzy environment. Knowl.-Based Syst. **61**(2), 48–58 (2014)

Decision Support and Behavior in Group Decision and Negotiation

Representative Decision-Making and the Propensity to Use Round and Sharp Numbers in Preference Specification

Gregory E. Kersten[1(✉)], Ewa Roszkowska[2],
and Tomasz Wachowicz[3]

[1] J. Molson School of Business, Concordia University, Montreal, Canada
gregory.kersten@concordia.ca
[2] Faculty of Economy and Management, University of Bialystok,
Bialystok, Poland
erosz@o2.pl
[3] Department of Operations Research, University of Economics in Katowice,
Katowice, Poland
tomasz.wachowicz@ue.katowice.pl

Abstract. This paper analyzes the agents' predisposition to produce round numbers during preference elicitation of the pre-negotiation phase. The agents negotiate on behalf of their principals and are asked to use information presented in terms of bar graphs and text to provide their principals' preferences numerically. In doing that, they tend to use round numbers more often than sharp numbers. Also, more agents use round numbers than sharp numbers, however, the majority of agents use a mix of numbers. The results show that the increased use of round numbers results in greater inaccuracy; the most accurate are agents who use a mix or round and sharp numbers.

Keywords: Preference reconstruction · Direct ratings assignment
Ratings accuracy · Heuristics · Round numbers · Sharp numbers

1 Introduction

Decision-makers use heuristics to process large amounts of information and to quickly determine solutions to complex problems. The result is that the decisions are often sub-optimal or even wrong. Making right decisions requires careful assessment of the decision problem, the problem's context and the preferences of the decision maker. Significant effort has been devoted to the design of procedures and algorithms that allow for the formulation of the problem and the preference system and the specification of one or more decision alternatives. In particular, numerous methods and aids were designed to elicit decision makers' preferences so that the obtained system correctly reflects their interests and meets predefined criteria [1].

In many cases the systems are used by agents, i.e., consultants, analysts, lawyers, and other support staff rather than the decision-makers (i.e., principals) themselves. This requires that the support staff know and accurately represent the principals'

Y. Chen et al. (Eds.): GDN 2018, LNBIP 315, pp. 43–55, 2018.
https://doi.org/10.1007/978-3-319-92874-6_4

preferences. A similar situation arises in representative negotiations when the negotiators are agents negotiating on behalf of the principals [2].

This paper raises the issue of the agents' ability regarding the formulation of a preference system that reflects the principals' preferences. There are several complementary approaches to study this issue. One is grounded in communication: the principals have to communicate their preferences. Communication and associated interpretation of messages have been discussed in [3, 4]. The second approach is based in extrinsic motivation; the principals need to motivate the agents so that the latter work diligently and represent the principals' true interests [5, 6]. The third approach is based on intrinsic motivation and the agents' innate abilities. Intrinsic motivation reflects the innate psychological needs for competence and self-determination and its impact on the cognitive effort and time devoted to the task. Both extrinsic and intrinsic motivations have been extensively studied [e.g., 5, 7, 8].

Motivation provides a reason to expend time and effort; strong motivation likely leads to using one's analytical skills and other qualities of the analytical rational information processing system, while weak motivation leads to the intuitive experiential system being used [9, 10]. The latter relies on heuristics that are often the cause of biases, albeit the analytical system may also produce biased answers due to the individuals' innate limitations and predispositions. Simon and Newell [11] and Tversky and Kahneman [12], conducted extensive research of the issues and published the results in their seminal works. Over 200 heuristics and cognitive biases were discovered [13]; for some their impact on the decision-making errors and pitfalls was assessed.

One heuristic that has not been researched in depth and, as far as we know, has not been studied in the context of preference elicitation, is the individuals' predisposition to use round numbers. In an early experiment, Kaufman et al. [14] showed sets of dots and observed that the participants had a strong tendency to produce round numbers, i.e., numbers divisible by 5 [cf. 15, pp. 97–111]. Jansen and Pollmann [16] analyzed texts corpora and noted that numbers divisible by five appear significantly more frequently, than the numbers in between (except for the number '2', which also appeared significantly more frequently than 0, 1, 3, and 4).

One difference in the perception of round and sharp numbers is due to the assumed effort in the number production [17]. Sharp numbers are seen as objective and accurate because they imply that they were obtained through analysis and computation. In contrast, round numbers being an approximation, are seen as subjective and imprecise estimation of reality.

If the round numbers are used as an approximation of precise numbers and are produced significantly more often than sharp numbers, then they may be used to construct the principals' preference system irrespectively of the principals' actual preferences. This possibility is studied here in the context of representative negotiations. We focus on the heuristic of using round and sharp numbers in the assignment of ratings used to construct a preference system. To this end, we conducted an experiment, a part of which required the agents reconstruct their principals' preferences based on both verbal and graphical description.

This work has implication for the construction of the preference system by the decision makers and their representatives. The heuristic of round numbers may lead to errors that may not be recognized in the elicitation models and procedures that are embedded in decision and negotiation support systems. The design of these systems should take into account the round number heuristic. Also, an inconsistency test may be required to determine whether the preferences formulated by the agent reflect the principal's preferences accurately.

The paper has five more sections. Section 2 discusses round numbers as focal points and the role of sharp and round numbers in decision-making. Section 3 presents four research questions and Sect. 4 briefly describes the negotiation experiments. The experiments' results and the use of round numbers in the construction of the scoring system are discussed in Sect. 5. Conclusions and discussion on future work are given in Sect. 6.

2 The Use and Interpretation of Round Numbers

The round number heuristic belongs to the category of focal point heuristics which directs or influences the decision maker' choices. Other heuristics in this category include anchoring, ideal points, and—in a more general sense—stereotypes. In the case of numeric representation these heuristics are used to simplify the representation to make it simpler and thus easier to think about [18, 19].

Focal point heuristics are used by individuals who need to assess an event or decide on a value associated with an incident, attribute, or characteristic. When they use rounded number (we call it RN heuristic), they are likely not to make an effort to determine the true value but round it to a number divisible by 5. RN allows to process numbers in an efficient and quick way. Their production is likely to be linked to the use of intuitive feeling-based information processing system rather than the employment of cognitively demanding analytical system. Conversely, processing of the sharp numbers (SN) is likely to rely on the analytical rational system that requires cognitive effort and analytical skills [20].

2.1 Frequency of Round and Sharp Number Use

Focal points are not necessarily RNs; for example, when people talk about time, focal numbers are 7 (days in a week), 24 (h in a day), 365 (days in a year), etc. If, however, they are required to give an uncertain value or choose a value from one of many, then they choose RN more often than SN [21].

Mason et al. [22] review of negotiation exercises shows that all opening offers made by 113 experienced executives and 243 MBA students were RN. The authors' review of 1511 offers made by sellers in real-estate markets shows that only 2% of the offers were SN. Under a reasonable assumption that the purpose of the opening offers is to establish an anchor and indicate willingness to make a concession, the RN offers need not be precise.

The tendency to use round numbers as approximations is clearly noticeable to the extent that when people know precisely that the value is a round number and communicate it they have to explain that the value is the exact one. Reporting a sharp (not round) number, however, does not need such an explanation [15, pp. 106–107].

2.2   RN and SN Interpretations

One may argue that the use of RN is rational because it conserves energy and effort. This may be the case in situations when RN and SN are equally likely to appear; such situations are, however, quite rare. The energy and effort conservation is a result of the use of intuitive experiential information processing system, therefore, communication of RN is likely to be perceived by the recipients differently than SN.

Communication of RN led the recipients to rely on the affective and intuitive dimensions (e.g., the relationship with the messengers and their appearance) while SN led the recipients to rely more on the cognitive and problem specific dimensions (e.g., the importance of the product attributes and budgetary implications). The difference between the perception of RN and SN has been observed in retail where prices are determined strategically; buyers who were not highly price conscious found such price as $19.99 substantially lower than $20.00, and $699 lower than $700 [17]. These buyers, however processed such prices not as SN but as RN, i.e., they considered $699 much lower than $700 to minimize cognitive effort required paying attention to the ending digit [23, 24].

RN and SN differently affect the adjustment of anchors. Janiszewski and Uy [25] conducted several lab and field experiments in which the participants were given RN and SN in 11 scenarios. The numbers established an anchor but the authors observed that the adjustment was different for RN and SN. They observed that a SN-based anchor required significantly less adjustment than a RN-based anchor, e.g., if the anchor was price, then the participants' estimation of price was significantly lower for RN-price than for SN-price [cf. 26].

In the negotiation context, the opening offers have been found to act as strong anchors and the negotiators are often biased in the direction of these offers [e.g., 27, 28]. Experiments show that SN-offers are more potent anchors than RN-offers because the former convey the offer-makers' confidence regarding the offer validity and true value. Negotiators who make SN-offers tend to be seen as more informed and reasoned than those who make RN-offers. RN-offers suggest that the offer-makers are less informed and less confident in the offer true value leading to a weaker anchoring effect [22].

RN are salient values; they are most likely to come to mind when individuals are trying to estimate a quantity that is uncertain. They are also likely to be used when the estimation would require effort that the individuals are not willing to extend. Therefore, the recipients of SN-type perceive their producers to be more persuasive and having greater competency than the producers of RN [29]. However, the recipients' knowledge of the subject described by SN and their skepticism regarding the SN producer reduces SN effect. On the other hand, if they trust the RN producer, then the difference between SN and RN assessment decreases.

To sum up, most of the earlier studies focused on the comparison of the frequency of RN and SN in different languages, text corpora, and decision processes and/or on the additional information that RN and SN conveyed about their producers. These results are useful here in that they show that different information processing systems tend to occur. Because we study here the use of RN and SN in the agent's re-construction of the principal's preferences, we are interested in the differences in the accuracy of the re-constructed preferences.

One study that deals with RN errors is the heaping measurement error, which is the error observed in the subjects' answers to survey's open-ended questions. Roberts and Brewer [30, p. 891] observed that all survey participants "all subjects who reported some large numerical estimates gave a 'round' number, and thus a multiple of 5, as a response." They tested two estimation methods: method C_1 checks the difference between any given response and the responses given by neighbors to the given response; and method C_2 considers only one neighbor at a time. They report that C_2 reduces the impact of the RN usage, the results, however, depend on the data set (op. cit. pp. 892–894).

Roberts and Brewer's [30, p. 891] approach is not directly applicable to our data set because the participants of our experiments are asked to use a verbal and graphical representation and make a numerical representation that is as accurate as it is possible. This means that the participants measurement error can be determined precisely.

3   Research Questions

Earlier studies suggest that people tend to describe uncertain, ambiguous and/or fluid situations using RN rather than SN. This lead us to formulate the following two research questions:

RQ1: Do the agents assign RN more frequently than SN to represent preferences of their principals, given that information about the principals' preferences is imprecise?

RQ2: Do more agents formulate RN-type preference systems rather than SN-type?

Many negotiators assume that their first offer would not be accepted, instead its purpose is to show their opening position and provide an anchor. Therefore, they tend to use round numbers; they may view a sharp number as associated with a pressure tactic indicating their inflexibility. The round numbers are perceived as intuitive while the sharp numbers are perceived as reasoned and based on careful analysis. The implication is that SN is the result of the rational analytical system (RAS) and RN are produced by experiential intuitive system (EIS). Thu, we may expect that production of RN to describe an object or phenomenon is likely to be associated with greater errors than SN production. This leads us to formulate the following research question:

RQ3: Are agents who construct their principals' preference system using RN more inaccurate than agents who use SN?

Given the relationship between RN and SN and the use of EIS and RAS, we also ask:

RQ4: Do agents who use RN tend to rely on EIS and do agents who use SN tend to rely on RAS?

4 Experimental Setup

To answer the above four questions, we use data from online bilateral negotiation experiments organized in 2015 and 2016 [31]. There were 984 students from the universities in Austria, Canada, Netherlands, Poland and Taiwan and a few from other universities.

The negotiation was between Fado, who represented a singer, and Mosico, a representative of an entertainment company. Both principals (singer and company) had their preferences regarding four negotiated issues; the four issues, options for each issue and the principals' preferences are given in Table 1.

Table 1. Principal's reference scoring systems for Mosico and Fado

Principal	Reference principal's ratings														
	No. of concerts				No. of songs				Royalties for CDs				Contract bonus		
	5	6	7	8	11	12	13	14	15	1.5	2.0	2.5	3.0	125	150	200
Singer	32	25	18	0	0	7	20	32	23	0	5	14	16	0	13	20
Company	0	23	31	39	0	6	16	30	24	10	20	15	0	11	5	0

The agents were not given the precise information shown in Table 1. Instead, the singer's and the company's preferences were communicated, respectively, to Fado and to Mosico in verbal and graphical format rather than numerical. An example of the information shown in Fig. 1 describes relative importance of four issues communicated to Fado. Similar information about the options for every issue was also given in the text and graph format. The experiment participants played the roles of Fado and Mosico and they were asked to use this information to reconstruct the principals' preferences as accurately as possible.

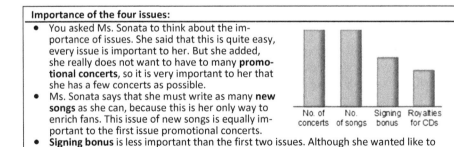

Importance of the four issues:
- You asked Ms. Sonata to think about the importance of issues. She said that this is quite easy, every issue is important to her. But she added, she really does not want to have to many **promotional concerts**, so it is very important to her that she has a few concerts as possible.
- Ms. Sonata says that she must write as many **new songs** as she can, because this is her only way to enrich fans. This issue of new songs is equally important to the first issue promotional concerts.
- **Signing bonus** is less important than the first two issues. Although she wanted like to make money, she must remain true to herself; that is, write and sign songs.
- She is the least concerned with the **royalties for CDs.**
- The illustration of the issue importance is given in the figure.

Fig. 1. An example of private information for Fado

The accurate scores for every issue are given in Table 1; they are the maximal values in each issue; i.e., No. of concerts and No. of songs are equally important with the weight of 32; Royalties has weight of 16; and Contract bonus' weigh is 20. The bars shown in Fig. 1 indicate the similarity and the differences in the weight values.

Comparing the preference ratings of Fado's and Mosico's principals we see that there are more SN in Fado's principal ratings than in Mosico's principal ratings. Because the use of RN and SN is context dependent, the consideration of both agents may produce different results. Therefore, to find answers to the four research questions we use ratings produced by Fado only.

5   Results

After eliminating incomplete records, the dataset consisted of 984 negotiators, among them 498 Fado agents. The students playing this role were from universities from Poland (40%), Austria (21.7%), China (8.2%), Taiwan (5%), Brazil (2.4%), France (1.6%), Canada (1.4%), Spain (1.4%), Finland (1.4%), USA (1.2%), Ukraine (1.2%) there were also students from 38 other countries represented with less than 1% from each country. They average age was 23.5 years and 57.4% of them were female. Their English proficiency was 4.77, understanding of the negotiation case 5.29 and negotiation experience 2.85 on a 1-7 Likert scale (1-low and 7-high).

The preference representation scheme required that in every issue the worst option had a rating equal zero. The participants were told that other options had greater rating value. Nonetheless, there were 13% (65 cases) of Fados who assigned rating of zero more than once in one or more issues. These participants were removed from the dataset and we further analyzed the remaining 433 cases.

5.1   Distribution of Issue Ratings

The histograms of ratings re-constructed by the 433 participants for every negotiation issue are shown in Figs. 2, 3, 4 and 5. The framed percent value (e.g., 5.8% in Fig. 2) is the percent of the participants who assigned exactly the same ratings as Fado's principal.

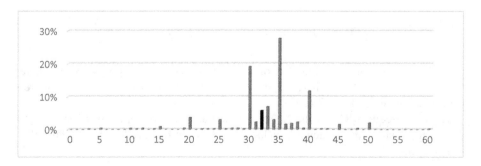

Fig. 2. The histogram of the ratings for issue *No. of concerts*

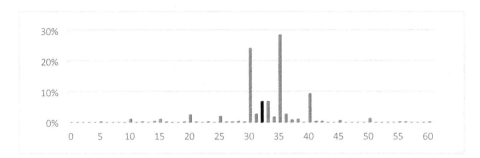

Fig. 3. Histogram of ratings for the issue *No. of songs*

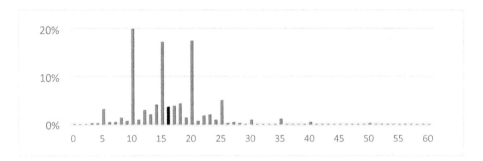

Fig. 4. Histogram of ratings for issue *Signing bonus*

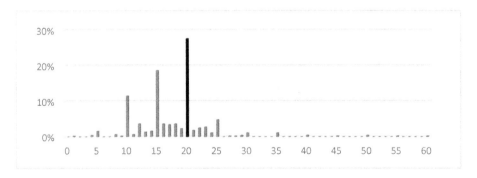

Fig. 5. Histogram of ratings for *Royalties issue*

For the No. of concert and the No. of songs, the two equally important issues, there are over 71% and 72% of RN, respectively. The distributions of these rating frequencies are given in Figs. 2 and 3.

The most frequent RN for the No. of concerts and for the No. of songs are 30, 35, 40. Rating equal to the principal's value, i.e., 32, was observed in 5.8% and 6.9% cases, respectively.

The histogram of the Signing bonus issue is similar to the two earlier issues in that only a small per cent (3.7%) of Fados selected SN equal to the principal's rating (16). Over 68% Fados selected RN, of which 60.6% selected 10, 15, 20, 25.

Royalties was the last and least important issue for the principal. The principal's rating for this issue was 20; 27.7% of Fados selected this rating. In total, 66.5% of Fados selected RN and 33.5% – SN (62.7% selected 10, 15, 20 and 25).

We obtained similar results for the options within each issue. Given that every participant was asked to select 12 numbers and there were 433 participants, results show that 61.8% of all the ratings are RN-type and 38.2% are SN-type. This means that the majority of ratings are of the RN-type. Thus, the answer to RQ1 is positive: the agents assign RN more frequently than SN.

5.2   Three Groups of Agents

In the group of 433 Fados, we have only 4 (0.9%) who assigned SN to every issue and option and 40 (9.2%) who assigned RN to every issue and option. Given that there were 12 options that should be assigned a non-zero value (the remaining 4 must had been assigned value 0), the remaining 389 Fados assigned between 1 and 11 RN.

For the purpose of further analysis, the agents were divided into three distinct groups:

1. RNG are agents who assigned at least 10 RN (no more than 2 SN);
2. SNG are agents who assigned at least 10 SN (no more than 2 RN);
3. MNG is group of agents who assigned between 3 and 9 RN or, equivalently, between 3 and 9 SN.

The number of agents in each group is as follows: RNG – 149; SNG – 33; and MNG – 252. This means that more agents formulated the principals' preferences using RN than SN, that is, the answer to RQ2 is positive.

5.3   Agents' Errors and Their Information Processing Systems

The accuracy of the agent's representation of the principal's preferences (see Table 1) is measured with the Hammond ordinal measure D_H and with the Manhattan (block) cardinal measure L_1 [see 32]. The comparison of the accuracy for the three groups is given in Table 2.

Alter et al. [20] observed that the production of RN is likely to rely on the intuitive experiential system while SN is likely to rely on the analytical rational system that requires cognitive effort. These two systems may operate in parallel or one of them is predominant. Epstein et al. [33] designed REI, a research instrument used to measure the strength of the individuals' rational and experiential processing systems.

The participants of the Inspire experiments were asked to answer questions included in REI-20. To determine the two systems RA (rational analytical) and intuitive experimental (EI), a confirmatory factor analysis with Varimax rotation and Kaiser normalization was conducted. The Kaiser–Meyer–Olkin (KMO) measure KMO = 0.888 indicates adequate sampling. KMO-values for the individual items which are greater than 0.802 are satisfactory [34]. The Bartlett's test indicated that the correlations between the items were sufficiently large to perform factor analysis [$\chi 2(190) = 4185.85$ p < 0.0001]. Both subscales RA and EI have high reliability with Cronbach's $\alpha = 0.898$ for Factor RA and $\alpha = 0.861$ for Factor EI. The overall variance explained is 50.46%. The values of RA and EI factors are given in Table 2.

Table 2. Agents' accuracy and their information processing systems

Agent group	No	L_I	D_H	Factor AR	Factor EI
Round Numbers (RNG)	149	67.70	3.08	−0.09	−0.02
All Round Numbers	40	71.17	4.70	−0.29	0.09
Mix Numbers (MNG)	251	64.38	2.78	0.08	0.01
Sharp Numbers (SNG)	33	58.06	2.48	−0.18	−0.01

Finally, a Mann-Whitney U test was used to analyze the differences between accuracy and information processing systems for pairs of groups. The results are presented in Table 3.

Table 3. Agents' scoring systems and REI

Test	L_I	D_H	Factor RA	Factor EI
RNG vs MNG	0.011^{**}	0.036^{*}	0.082^{*}	0.887
MNG vs SNG	0.080^{*}	0.333	0.304	0.896
RNG vs SNG	0.002^{***}	0.049^{*}	0.969	0.800
ARN vs SNG	0.006^{***}	0.000^{***}	0.542	0.965

$^{*}p < 0.1;$ $^{**}p < 0.05;$ $^{***}p < 0.01$

Using ordinal and cardinal measures we found that among the three groups the RNG group is significantly less accurate than MNG and SNG. The accuracy of ARN (all round numbers) subgroup of RNG is even lower; ARN is the least accurate. The most accurate Fados belong to the group that uses both round and sharp numbers, i.e., MNG.

From Table 3 we obtain the answer to RQ3: agents who construct their principals' preference system using RN are more inaccurate than agents who use SN. We do not find, however, the confirmation to RQ4 that agents who use RN tend to rely on experiential intuitive system, while agents who use SN tend to rely on rational analytical system. We find no significant ($p \geq 0.800$) difference between groups with respect to EIS. There is only weakly significant difference ($p < 0.1$) between RNG and MNG with respect to RAS. The agents from the group MNG are more analytical-rational than those from the group ARN or RNG.

6 Conclusions

We have identified a tendency to use round numbers in the reconstruction of the principals' preference values. The results confirm earlier observations that there are spikes in the distribution at round numbers and round numbers are used more often than sharp numbers. We also confirmed that the use of round numbers is associated with greater errors. Members of the group who produced at least 10 RN out of 12 numbers made greater ordinal and cardinal errors than the members who produced fewer than 10 RN. The most accurate group was the one that used no more than 2 RN and not fewer than 10 SN.

To some extent, these results are context dependent. The communication of the accurate numbers, i.e., the principal's preferences, was indicative and inaccurate so that the agents were likely to make cardinal errors. However, they should not have made ordinal errors because the bars show the order unequivocally when their height difference is measured with natural numbers. We observed that the predisposition to use round numbers increases ordinal error.

One implication of this study is that the designers of preference elicitation tools may consider using linguistic or fuzzy scales rather than real or natural number scales. We need to stress, however that, as far as we know, this is the first study on the use and implication of round numbers in preference elicitation. In order to make better informed recommendations, more studies are required.

We were unable to confirm the relationship between information processing systems used by the experiments' participants and their tendency to produce round and sharp numbers. It would appear that the use of the information processing type does not depend on context and task. We found, however, no significant relationship between rational analytical system and intuitive experiential system and the production of sharp and round numbers. One possibility is that REI-20 instrument does not allow to assess the strength of the two systems in individuals from different cultures or in the context in which we used it. This issue should be studied further because decision and negotiation support systems ought to be tailored to the abilities and predispositions of their users.

Acknowledgements. This research was supported with the grants from Polish National Science Centre (2016/21/B/HS4/01583) and from the Natural Sciences and Engineering Canada.

References

1. Aloysius, J.A., et al.: User acceptance of multi-criteria decision support systems: the impact of preference elicitation techniques. EJOR **169**(1), 273–285 (2006)
2. Rubin, J.Z., Sander, F.E.: When should we use agents? Direct vs. representative negotiation. Negot. J. **4**(4), 395–401 (1988)
3. Dessein, W.: Authority and communication in organizations. Rev. Econ. Stud. **69**(4), 811–838 (2002)
4. Solet, D.J., et al.: Lost in translation: challenges and opportunities in physician-to-physician communication during patient handoffs. Acad. Med. **80**(12), 1094–1099 (2005)

5. Deci, E.L., Koestner, R., Ryan, R.M.: Extrinsic rewards and intrinsic motivation in education: reconsidered once again. Rev. Educ. Res. **71**(1), 1–27 (2001)
6. Laffont, J.J., Martimort, D.: The Theory of Incentives: The Principal-Agent Model. Princeton University Press, Princeton (2002)
7. Bénabou, R., Tirole, J.: Incentives and prosocial behavior. Am. Econ. Rev. **96**(5), 1652–1678 (2006)
8. Eisenberger, R., Pierce, W.D., Cameron, J.: Effects of reward on intrinsic motivation: negative, neutral, and positive: comment on Deci, Koestner, and Ryan (1999). Psychol. Bull. **125**(6), 677–691 (1999)
9. Chaiken, S., Trope, Y.: Dual-Process Theories in Social Psychology. Guilford Press, New York (1999)
10. Epstein, S.: Cognitive-experiential self-theory of personality. In: Millon, T., Lerner, M. J. (eds.) Handbook of Psychology. Wiley, Hoboken (2003)
11. Simon, H.A., Newell, A.: Heuristic problem solving: the next advance in operations research. Oper. Res. **6**(1), 1–10 (1958)
12. Tversky, A., Kahneman, D.: The framing of decisions and the psychology of choice. Science **211**(4481), 453–458 (1981)
13. Benson, B.: Cognitive bias cheat sheet. Better Humans 2016. https://betterhumans.coach.me/cognitive-bias-cheat-sheet-55a472476b18. Accessed 4 Jan 2017
14. Kaufman, E.L., et al.: The discrimination of visual number. Am. J. Psychol. **62**(4), 498–525 (1949)
15. Dehaene, S.: The Number Sense: How the Mind Creates Mathematics. Oxford University Press, Oxford (2011)
16. Jansen, C.J., Pollmann, M.M.: On round numbers: pragmatic aspects of numerical expressions. J. Quant. Ling. **8**(3), 187–201 (2001)
17. Schindler, R., Yalch, R.: It seems factual, but is it? Effects of using sharp versus round numbers in advertising claims. In: Pechmann, C., Price, L. (eds.) ACR North American Advances, Association for Consumer Research: Duluth, MN, pp. 586–590 (2006)
18. Gilovich, T., Griffin, D., Kahneman, D.: Heuristics and biases: the psychology of intuitive judgment. Cambridge University Press, Cambridge (2002)
19. Schelling, T.C.: The Strategy of Conflict. Harvard University Press, Boston (1980)
20. Alter, A.L., et al.: Overcoming intuition: metacognitive difficulty activates analytic reasoning. J. Exp. Psychol. Gener. **136**(4), 569 (2007)
21. Baird, J.C., Lewis, C., Romer, D.: Relative frequencies of numerical responses in ratio estimation. Attention Percept. Psychophy. **8**(5), 358–362 (1970)
22. Mason, M.F., et al.: Precise offers are potent anchors: conciliatory counteroffers and attributions of knowledge in negotiations. J. Exp. Soc. Psychol. **49**(4), 759–763 (2013)
23. Bizer, G.Y., Schindler, R.M.: Direct evidence of ending-digit drop-off in price information processing. Psychol. Mark. **22**(10), 771–783 (2005)
24. Thomas, M., Morwitz, V.: Penny wise and pound foolish: the left-digit effect in price cognition. J. Consum. Res. **32**(1), 54–64 (2005)
25. Janiszewski, C., Uy, D.: Precision of the anchor influences the amount of adjustment. Psychol. Sci. **19**(2), 121–127 (2008)
26. Pope, D.G., Pope, J.C., Sydnor, J.R.: Focal points and bargaining in housing markets. Games Econ. Behav. **93**, 89–107 (2015)
27. Galinsky, A.D., Mussweiler, T.: First offers as anchors: the role of perspective-taking and negotiator focus. J. Pers. Soc. Psychol. **81**(4), 657–669 (2001)
28. Bazerman, M.H., Neale, M.: Heuristics in negotiation: limitations to dispute resolution effectiveness. In: Bazerman, M.H., Lewicki, R.J. (eds.) Negotiations in Organizations, pp. 51–67. Sage, Beverly Hills (1983)

29. Xie, G.-X., Kronrod, A.: Is the devil in the details? The signaling effect of numerical precision in environmental advertising claims. J. Advertising **41**(4), 103–117 (2012)
30. Roberts, J.M., Brewer, D.D.: Measures and tests of heaping in discrete quantitative distributions. J. Appl. Stat. **28**(7), 887–896 (2001)
31. Kersten, G., Roszkowska, E., Wachowicz, T.: An impact of negotiation profiles on the accuracy of negotiation offer scoring system - experimental study. Multiple Criteria Decis. Making **11**, 77–95 (2016)
32. Roszkowska, E., Wachowicz, T.: Inaccuracy in defining preferences by the electronic negotiation system users. In: International Conference on Group Decision and Negotiation, pp. 131–143. Springer, Heidelberg (2015).https://doi.org/10.1007/978-3-319-19515-5_11
33. Epstein, S., et al.: Individual differences in intuitive–experiential and analytical–rational thinking styles. J. Pers. Soc. Psychol. **71**(2), 390 (1996)
34. Hutcheson, G., Sofroniou, N.: The Multivariate Social Scientist. Introductory Statistics Using Generalized Linear Models. Sage, London (1999)

Neuroscience Experiment for Graphical Visualization in the FITradeoff Decision Support System

Lucia Reis Peixoto Roselli, Eduarda Asfora Frej[(⊠)],
and Adiel Teixeira de Almeida

CDSID - Center for Decision Systems and Information Development,
Federal University of Pernambuco – UFPE, Av. Acadêmico Hélio Ramos,
s/n – Cidade Universitária, Recife, PE 50740-530, Brazil
luciarpr@hotmail.com, {eafrej,almeida}@cdsid.org.br

Abstract. The neuroscience approach is considered to be a study of the neural system and its implications for processes in the human body. Behavioral studies in Multicriteria Decision Making (MCDM) still have a gap and in this context, Neuroscience can be used as a decision support tool. Therefore, the aim of this research study is to explore the potential of using graphical visualization in the FITradeoff Decision Support System (DSS) by undertaking an eye-tracking experiment and applying it to a decision problem. In the end, based on the results, suggestions are made to the analyst and improvements are made to the design of the DSS so that solutions could be found that accurately express a decision maker's preferences.

Keywords: Neuroscience · Multicriteria decision-making · FITradeoff
Eye-tracking

1 Introduction

The human brain is the most complex organ in the human body. Therefore, with a view to reaching a better understanding of how the brain functions, the Neuroscience approach was developed. Neuroscience engages on the study of neural system and promotes understanding of how the mechanisms of our body function. Neuroscience has been used by many areas of knowledge to improve systems [23].

With regard to decision making, this approach seeks to provide a fuller understanding of the mechanisms that underlie the decision process. As Neuroscience can be related to the decision process of many different areas, some specific approaches have been developed, such as: Neuroeconomics, NeuroIS, Consumer Neuroscience, Neuromarketing, Management Neuroscience and Organizational Neuroscience [20].

Neuroeconomics has become a complement to classical economic theories, since these alone are no longer sufficiently broad to represent and fully encompass the decision process [3, 4, 12, 15]. NeuroIS was developed to better understand cognition, emotion and behavior processes and arose from research studies on neuro-adaptive information systems [16].

Consumer Neuroscience is used to identify consumers' preferences, while Neuro-marketing leads to products that are compatible with consumers' preferences. These approaches have been developing suggestions to guide design concepts and to present products [5, 9, 13].

Due to neuroscience having become an important support tool for several areas of knowledge, several kinds of equipment that measure body variables have been developed. These include: galvanic skin response sensors, heart rate meters; and devices that measure electric signals between neurons, the oxygenation rate of hemoglobin molecules, and ocular movements.

In this context, experiments have been developed using tools to analyze some decision situations. Using fMRI (functional Magnetic Resonance Imaging) to analyze brain activation, Sanfey et al. [17] presented a simple game, called Ultimatum Game, to evaluate the limitations of classical economic models in providing a real representation of the decision-making process. Goucher-Lambert et al. [5] and Sylcott et al. [21] evaluate consumers' preference judgments for sustainable products and the combination of the form and function of a product.

With specific regard to eye movements, using eye-tracking, Ares et al. [1] and Guixeres [6] evaluate the differences between yogurt labels and the effectiveness of ads. Using eye-tracking and electroencephalograph (EEG), Slanzi et al. [19] and Khushaba [9] evaluate clicks on five websites and consumers' preference for three types of crackers.

As to the multicriteria decision process and Neuroscience, there are papers in the literature that evaluate several criteria but none of them use Neuroscience as tool to support multicriteria decision processes, showing the gap between these approaches [7, 10].

Therefore, this paper sets out to evaluate behavioral aspects in the FITradeoff method. To do so, an experiment was undertaken and results evaluated. This experiment was developed to analyze the specific step of graphical visualization in the FITradeoff Decision Support System (DDS). Thus, the research question concerns how decision makers evaluate graphical visualization and, therefore, how does this lead them to select the best alternative. To conduct this experiment, eye-tracking equipment was used. There were two end-purposes: to give insights to the analyst and to improve FITradeoff DSS.

This paper is organized as follows. Section 2 gives a brief description of the FITradeoff Method. Section 3 describes a behavioral experiment; Sect. 4 gives the results from the experiment while Sect. 5 analyzes and discusses these results. Final remarks are made and some conclusions are drawn in Sect. 6, which also suggests some lines for future research studies.

2 Flexible Interactive Tradeoff Method

The Flexible Interactive Tradeoff method - FITradeoff [2], was developed in order to elicit scaling constants in the context of Multi-Attribute Value Theory – MAVT [8]. This method is based on the Traditional Tradeoff [8] which has the same axiomatic structure, but FITradeoff has some advantages when compared with the traditional method.

The FITradeoff method has three steps, which seek to evaluate the intra-criteria utilities, to rank the criteria weights and to evaluate the criteria weights. The first step is common to most multicriteria methods and in this case, the decision maker (DM) imports the decision matrix.

The second step is the same as in the Traditional Tradeoff, namely, the DM compares the criteria weights and ranks these criteria. After this step, the first inequality is obtained, presented in expression (1), in which k_i is the scaling constant of criterion i.

$$k_i > k_j > k_m...k_n \tag{1}$$

The third step is characterized as being when consequences are compared in the decision matrix. Thus, adjacent criteria are compared. The best consequence of the second criterion is compared to a hypothetical consequence of the first criterion, which is lower than the best consequence of the first criterion. So, from a relation with the strict preference expressed by the DM, two inequalities can be obtained, as shown in expressions (2) and (3).

$$k_j v_j(x_j') > k_{j+1} \tag{2}$$

$$k_j v_j(x_j'') < k_{j+1} \tag{3}$$

Compared to the Traditional Tradeoff elicitation procedure, the difference between this step and the original one is the absence of the indifference point. In FITradeoff, the DM does not need to express the exact point of indifference, which is why it is considered to be cognitively easier to understand. According to Weber and Borcherding [22], difficulties found in identifying indifference points leads to 67% inconsistency in results.

After each comparison has been made by the DM, a linear programing problem (LPP) is solved using the inequalities obtained above. These inequalities represent the DM's preference for one or other consequence. Thus, after each LPP has been solved, the range of initial alternatives decreases and they become Potentially Optimal Alternatives (POA). FITradeoff is considered interactive because of this step, where the DM makes comparisons and analyzes POA throughout the whole process.

Another advantage of FITradeoff is that it presents information that can be visualized graphically, in particular POAs, which helps the DM to make decisions. This feature characterizes FITradeoff as being a flexible and important tool because the time that the DM takes to process information is reduced and consequently, the DM can reach the final solution more quickly and therefore stop the process of seeking the best alternative.

Besides making use of a Neuroscience approach, the focus of this research is on analyzing ways to use graphical visualization. The next section discusses the three types of graphical visualization supported by the FITradeoff DSS (Bar Graph, Bubble Graph and Spider Graph – see Fig. 1) and other two added. The FITradeoff elicitation process is illustrated in Fig. 2.

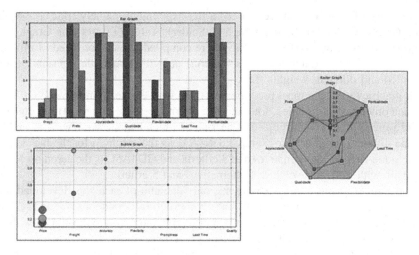

Fig. 1. FITradeoff graphics

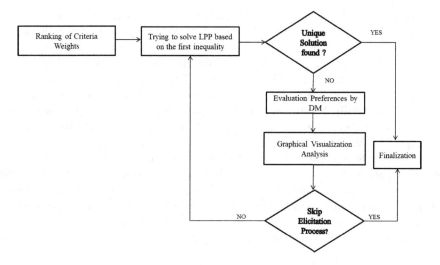

Fig. 2. FITradeoff process

The FITradeoff method is available by request to the authors at www.fitradeoff.org.

3   Behavioral Experiment with Neuroscience Tools

Graphical Visualization can be present in a decision-making process as a support tool to help the DM. Therefore, an experiment was undertaken, the purpose of which was to analyze how the DMs both understood different types of graphic visualization and used them to make decisions.

Five types of visualization were used in the experiment, namely: Bar Graph (G), Bubble Graph (GBubble), Spider Graph (GSpider), Table (T) and Bar Graph with Table (GT). In total, twenty-four graphics were compiled, which consisted of different combinations of items (alternatives vs. criteria) and different scale constants (same weights (S) and different weights (D)).

Bar graphics were the most predominant type, with eighteen units, which differed from each other by having three, four and five alternatives and criteria. These eighteen graphics were split into two groups of nine, one of which had the same weights and the other had different weights. For example, GS3A3C is the acronym for the bar graphic with the same weights, 3 alternatives and 3 criteria and GD4A5C is the acronym for the bar graphic with different weights, 4 alternatives and 5 criteria.

As to Bubble graphs and Spider graphs, only one unit of each was developed with the same weights, four alternatives and five criteria (GSpider4A5C and GBubble4A5C). For the Table and Bar Graph with Table, two units were developed with the same weights, three or four alternatives and five criteria (T3A5C, T4A5C, GT3A5C and GT4A5C). These types of visualization were developed to compare with the corresponding bar graphics (GS3A5C and GS4A5C) aiming to analyze which is the best for the DM.

After the graphics had been developed, they were mixed into three distinct sequences. The first sequence, called S1, was characterized by the growth in the degree of difficulty for the DM related to the number of items. S1 was developed with nine bar graphs, with the same weights, followed by the six others types of visualization and finally nine bar graphs with different weights. The second sequence, S2, had the characteristic of decreasing degree of difficulty and was constructed in the opposite way to S1. And finally, S3 presented the bar charts in a totally random way. In general, sequences had twenty-four visualization shapes, varying the position of the bar graphs and keeping the different visualization shapes, in the middle of the sequence.

To conduct the experiment the eye-tracking equipment X120 by Tobbi Studio was used. This equipment uses emission of infrared rays and the reflection of these by the cornea to measure the eye movements. Based on elements present in the eye-tracking software, the three similar experiments, each of which had one sequence, comprised: explanatory slides, images of each form of visualization and questionnaires. The questionnaires were presented after each image and had the following question: What is the best alternative?

The best alternative was previously defined by the researcher using the Additive Model [19]. The researcher wished to evaluate the hit rate (HR) for each graph and how it interacted with the left eye pupil diameter (LEPD) and fixation duration (FD).

An initial sample of fifty-four management engineering students and PhD professors took part in the experiment. A total of thirty-six recordings of eye movements were used and the results from these were analyzed. The recordings were of sixteen undergraduate students, ten master's degree students, six doctoral students and four PhD Professors. The sequence in which each participant took part in the experiment was determined at the convenience of the researcher and in accordance with the availability of the participants. There was a sample population of twelve participants for each of the sequences.

Finally, meetings were held in the NSID (NeuroScience for Information and Decision) laboratory. Prior instructions were provided in same way for each participant and the research project was approved by the Ethics Committee of the Federal University of Pernambuco before the data were collected. Figure 3 shows a participant taking part in a real experiment.

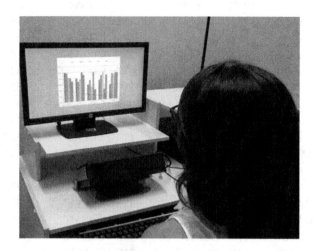

Fig. 3. A participant in the experiment

4 Results of the Experiment

Based on variables collected in the experiment, some results were generated in order to analyze the HR variable. This was considered the most relevant variable because of its relationship to the research question: how do participants understand the graphical visualization forms and, therefore, how does this lead them to select the best alternative?

Hit Rate values were derived from the ratio of the number of correct answers to the total number of answers for each graph. A correct answer was deemed to be the best alternative for each graph, previously found by the Additive Model. Therefore, the researcher compared the participants' answers to the questionnaires with the answers from the Additive Model to determine how many of a participant's answers were correct. Table 1 presents the HR for each type of visualization, following the order that it appears in each sequence.

Table 1. Hit rate

S1	HR	S2	HR	S3	HR
GS3A3C	83%	GD5A5C	42%	GS3A4C	92%
GS4A3C	25%	GD4A5C	58%	GD4A5C	67%
GS5A3C	58%	GD3A5C	25%	GS3A5C	33%
GS3A4C	92%	GD5A4C	75%	GD5A4C	75%
GS4A4C	50%	GD4A4C	8%	GS4A4C	75%
GS5A4C	75%	GD3A4C	25%	GD4A3C	50%
GS3A5C	17%	GD5A3C	25%	GD3A4C	8%
GS4A5C	50%	GD4A3C	67%	GD5A3C	33%
GS5A5C	75%	GD3A3C	33%	GD3A5C	33%
GSpider4A5C	75%	GSpider4A5C	100%	GSpider4A5C	92%
GBubble4A5C	42%	GBubble4A5C	58%	GBubble4A5C	50%
T3A5C	25%	T3A5C	17%	T3A5C	33%
T4A5C	83%	T4A5C	92%	T4A5C	75%
GT3A5C	25%	GT3A5C	8%	GT3A5C	8%
GT4A5C	50%	GT4A5C	75%	GT4A5C	75%
GD3A3C	42%	GS5A5C	58%	GS5A4C	67%
GD4A3C	58%	GS4A5C	75%	GD5A5C	58%
GD5A3C	17%	GS3A5C	8%	GS3A3C	100%
GD3A4C	25%	GS5A4C	75%	GS4A3C	50%
GD4A4C	8%	GS4A4C	67%	GS5A5C	92%
GD5A4C	83%	GS3A4C	100%	GD3A3C	75%
GD3A5C	17%	GS5A3C	58%	GD4A4C	17%
GD4A5C	42%	GS4A3C	33%	GS4A5C	92%
GD5A5C	42%	GS3A3C	100%	GS5A3C	75%

The first analysis was developed in an attempt to explain HR using the FD, based on the reasoning: the longer that a graph is visualized, the more difficult it is to analyze it and the lower the HR value. Thus, the following research question was: Is there a correlation between the variables (FD and HR) for each sequence?

In order to develop this analysis, the original FD values that had been extracted from each of the recordings, were simplified. The average of the twelve values extracted for each graph, was calculated, resulting in a final value for each graph, as shown in Table 2.

Table 2. Average duration of fixation in milliseconds

Graphics	FD for S1	FD for S2	FD for S3
GS3A3C	11.75	11.51	15.22
GS4A3C	13.67	11.68	17.78
GS5A3C	14.3	16.46	21.77
GS3A4C	14.42	19.36	21.78
GS4A4C	16.04	24.18	22.56
GS5A4C	17.02	25.11	24.72
GS3A5C	17.22	26.04	25.37
GS4A5C	20.14	27.31	27.11
GS5A5C	20.5	28.01	28
GD3A3C	20.53	29.01	28.89
GD4A3C	22.02	29.46	29.1
GD5A3C	23.94	29.85	29.88
GD3A4C	24.85	30.04	30.11
GD4A4C	25.39	30.13	30.35
GD5A4C	25.82	30.46	31.02
GD3A5C	27.53	31.08	31.52
GD4A5C	28.77	31.65	31.54
GD5A5C	29.28	35.2	35.06
GSpider4A5C	29.38	35.84	35.5
GBubble4A5C	31.82	37.89	40.07
T3A5C	32.29	42.83	41.16
T4A5	37.14	44.66	46.29
GT3A5C	38.3	45.06	53.68
GT3A5C	44.52	59.41	54.87

The second analysis was developed to evaluate the HR using the LEPD. As to pupil diameter, several studies have proven that this variable has a strong relationship with the intensity of mental activity: the diameter is greater when a greater effort is made [11, 14].

Thus, this analysis is based on the reasoning: the larger pupil diameter is, the more difficult it is for someone to analyze a visual and therefore the lower the HR value is. A similar research question was drawn up to test this hypothesis: Is there a correlation between the variables (LEPD and HR) for each sequence?

In order to perform this analysis, all LEPD values captured during the recordings were extracted and separated into each visualization type for each participant. Thereafter, an average was calculated for each participant and thus a single value of the LEPD in each visualization type was obtained. Finally, another average was calculated of the twelve values of LEPD, found for each participant and thus, a unique value for each graphic in each sequence was obtained. For all visualization shapes, the LEPD ranged between 4.05 and 4.75 mm.

Only the left eye was chosen for the analysis so as to simplify the research experiment. Choosing to do so is supported by the literature which gives evidence that the results from analyzing either eye are indifferent to each other [18].

For these two analyses, the Spearman Correlation was applied to find the relationship between the variables selected for evaluation. The results are given in Table 3. The absence of a strong causality between these variable was evidenced by the low correlation rates.

Table 3. Results of the Spearman Correlation

Variables	S1	S2	S3
FD: HR	−0.13	−0.29	0.01
LEPD: HR	0.32	0.24	−0.01

Because of the absence of correlation in the analysis above, a final descriptive analysis was developed with a view to recommending a minimum confidence level for graphs with the same number of items. The aim of this descriptive analysis was to support the analyst in his recommendations on whether or not to use graphical visualization in decision problems.

To perform this analysis, a quality interval was constructed using percentage of acceptance levels estimated by the researcher based on the amount of wrong answers in each graph. This interval was built using levels of acceptance on the number of wrong answers in each graph, as shown in Table 4.

Table 4. Quality interval

Percentage	Maximum number of errors	Classification
p1 < 0.2	2.4	VG - Very Good
p2 < 0.3	3.6	G - Good
p3 < 0,5	6	R - Satisfactory
p4 < 0.6	7.2	D - Unsatisfactory

Therefore, from the aggregation of the HR and the classification for each graphic, the minimum confidence level was estimated based mainly on worst values of HR in S3. This category was chosen due to the randomness of S3 thus trying to find a more assertive level of confidence for all graphics. The confidence level for six types of visualization, comparing with the corresponding bar graph, is shown in Table 5. For the bar graph, those with equal weights are compared to those with different weights, as shown in Table 6.

Table 5. Confidence level for six other types of visualization

Graphics	S1	S2	S3	Level of confidence
GS3A5C	0.17	0.08	0.33	10%
Classification	D	D	D	
T3A5C	0.25	0.17	0.33	
Classification	D	D	D	
GT3A5C	0.25	0.08	0.08	
Classification	D	D	D	
GS4A5C	0.50	0.75	0.92	75%
Classification	R	G	VG	
GSpider4A5C	0.75	1.00	0.92	
Classification	G	VG	VG	
GBubble4A5C	0.42	0.58	0.50	
Classification	D	R	R	
T4A5C	0,83	0.92	0.75	
Classification	VG	VG	G	
GT4A5C	0.50	0.75	0.75	
Classification	R	G	G	

Table 6. Confidence level for bar graphs

Graphics	S1	S2	S3	Level of confidence
GS3A3C	0.83	0.42	0.92	75%
Classification	VG	VG	VG	
GD3A3C	0.42	0.33	0.75	
Classification	D	D	R	
GS4A3C	0.25	0.33	0.50	50%
Classification	D	D	R	
GD4A3C	0.58	0.67	0.50	
Classification	R	R	R	
GS5A3C	0.58	0.58	0.75	30%
Classification	R	R	G	
GD5A3C	0.17	0.25	0.33	
Classification	D	D	D	
GS3A4C	0.92	1.00	0.92	30%
Classification	VG	VG	VG	
GD3A4C	0.25	0.25	0.08	
Classification	D	D	D	
GS4A4C	0.50	0.67	0.75	20%
Classification	R	R	G	
GD4A4C	0.08	0.08	0.17	
Classification	D	D	D	

(*continued*)

Table 6. (*continued*)

Graphics	S1	S2	S3	Level of confidence
GS5A4C	0.75	0.75	0.67	75%
Classification	G	G	R	
GD5A4C	0.83	0.75	0.75	
Classification	VG	G	G	
GS3A5C	0.17	0.08	0.33	30%
Classification	D	D	D	
GD3A5C	0.17	0.25	0.33	
Classification	D	D	D	
GS4A5C	0.50	0.75	0.92	70%
Classification	R	G	VG	
GD4A5C	0.42	0.58	0.67	
Classification	D	R	R	
GS5A5C	0.75	0.58	0.92	60%
Classification	G	R	VG	
GD5A5C	0.42	0.42	0.58	
Classification	D	D	R	

In addition to the analyses of the HR, another complementary analysis was developed using areas of interest (AOI) to further enhance the FITradeoff DSS. Areas of interest were regions drawn in each graphic to collect the variables. Based on eye-tracking, the FD was collected for each graph and for each specific area within the graphs. Thus, areas of interest were set for each criterion, in each bar graphics with different weight, in order to evaluate how participants visualized each criterion and with a view to confirming that weights were being positioned consistently (from left to right) in the FITradeoff DSS. The Table 7 shows the criterion most visualized in each graphic for each sequence.

Table 7. AOI analysis

Graphics	Decreasing weights S1		Decreasing weights S2		Decreasing weights S3	
	Most looked AOI	Second most looked AOI	Most looked AOI	Second most looked AOI	Most looked AOI	Second most looked AOI
G3A3C	Criterion 2	Criterion 1	Criterion 2	Criterion 1	Criterion 2	Criterion 1
G4A3C	Criterion 2	Criterion 1	Criterion 1	Criterion 2	Criterion 2	Criterion 1
G5A3C	Criterion 2	Criterion 1	Criterion 2	Criterion 1	Criterion 2	Criterion 1
G3A4C	Criterion 2	Criterion 1	Criterion 2	Criterion 1	Criterion 2	Criterion 1
G4A4C	Criterion 1	Criterion 2	Criterion 2	Criterion 1	Criterion 2	Criterion 3
G5A4C	Criterion 2	Criterion 1	Criterion 1	Criterion 2	Criterion 2	Criterion 3
G3A5C	Criterion 2	Criterion 1	Criterion 2	Criterion 1	Criterion 1	Criterion 2
G4A5C	Criterion 1	Criterion 2	Criterion 2	Criterion 1	Criterion 1	Criterion 2
G5A5C	Criterion 1	Criterion 2	Criterion 1	Criterion 2	Criterion 1	Criterion 2

Thus, based on Table 7, for graphs different weights, the left and central criteria were the most visualized since they received the highest FD. This result proves that criteria were properly positioned in the FITradeoff DSS. The next section offers further comments on the results developed.

5 Discussion of Results

The goal of the experiment was to evaluate the graphic visualization, given the flexibility that this tool brings to FITradeoff method. In addition, based on the research question – how do decision makers evaluate graphical visualization and how do they use this to select one of the final alternatives? There were two main purposes – to evaluate whether or not the graphic visualizations both aid the analyst to have insights and improve the FITradeoff DSS. The analyses that were developed are discussed in this section.

The Hit Rate variable was focused on in this research due to its relevance for the research question. Therefore, two variables (FD and LEPD) were collected using eye-tracking and the Spearman Correlation (HR v FD and HR v LEPD) was calculated in an attempt to explain the HR.

However, based on the results of the correlation, it was not possible to verify the relationship between these variables. So, it was not possible to state anything about the difficulty related to the number of items, based on FD and LEPD. To further explore the HR, a descriptive analysis was performed, thereby providing a confidence level for each type of visualization which had a similar number of items.

Therefore, based on the confidence level, some conclusions can be supposed. First, S3 had a greater number of hits than the other two sequences. Secondly, the bar graphs with equal weights had a higher HR when compared individually with graphs with different weights. Thirdly, the Spider graph may be more appropriate for problems with a large number of items and, fourthly, in this experiment, the Tables received a higher HR compared to the other types of visualization.

Based on these conclusions and the two main goals of the experiment, give insights to the analyst and improve FITradeoff DSS, it is observed that: the conclusions about bar graphs with equal weights and the spider graph can be used by the analyst, in addition to Table 2, as possibilities that may well assist solving decision problems. And the fourth conclusion that tables receive a higher HR than other forms of graphical visualization can be used as a recommendation to include the possibility of using them in FITradeoff DSS, since they have not been included already. As to enhancing FITradeoff DSS, based on AOI analysis, this form of analysis confirmed that when criteria weights are placed in left-central positions, they received the highest FD and consequently highest weights.

More generally, in addition to discussions about results, further questions can also be generated. Thus, this research study leads to the need to ask further questions, such as: If the sample is most diverse, would tables continue to lead to better HR? Were other factors maybe associated with HR? This factor can be the way of how data was composed (indicating that some decision matrices should not be used to build graphics)? These questions can be explored in future research.

6   Conclusion

Neuroscience approach is characterized as a study of the neural system and how this affects processes in the human body. Therefore, this research study was developed with a view to integrating neuroscience into a multicriteria decision-making approach, and in particular, for the FITradeoff Method.

Several studies in the literature have applied neuroscience experiments to decision-making, but most of them are related to decision in cognitive sense within a health context. As regard to decision-making in the organizational context, most of these studies are related to risk decision analysis, many of them within utility theory background. Yet, no studies have been found in the literature which do so in conjunction with specific multicriteria methods, as it is the case of either tradeoff elicitation procedure or FITradeoff method. Thus, this paper has developed and applied an experiment, using eye-tracking, to investigate how the participants understood graphical visualization and if this led them to selecting the best alternative. The objectives were to improve the design of the DSS and to assist the analyst in obtaining insights.

As to the first objective, a confidence level for each type of visualization was developed and can be used in any other multicriteria method at the discretion of the analyst. This possibility is particularly helpful if the analyst has not been hitherto aware of what graphical visualization shapes may assist in tackling decision problems. With regard to the second objective, this paper shows that preliminary studies indicate that the use of tables led the participants to better answers than other visuals. Currently, the DSS does not include the possibility of using tables. Therefore, this paper suggests that including tables in the DSS may well assist a DM. Further studies need to be conduct on this topic and also to consider the possibility that this may depend on different DM's styles.

Finally, as suggestions for future research studies, the authors recommend developing experiments to investigate issues in eliciting preferences which is related to Step three of the FITradeoff method. Secondly, it would be helpful to replicate the experiment undertaken in this study with a larger and more diverse sample population.

Acknowledgments. This study was partially sponsored by the Brazilian Research Council (CNPq) for which the authors are most grateful.

References

1. Ares, G., et al.: Influence of rational and intuitive thinking styles on food choice: preliminary evidence from an eye-tracking study with yogurt labels. Food Qual. Prefer. **31**, 28–37 (2014)
2. de Almeida, A.T., De Almeida, J.A., Costa, A.P.C.S., De Almeida-Filho, A.T.: A new method for elicitation of criteria weights in additive models: flexible and interactive tradeoff. Eur. J. Oper. Res. **250**, 179–191 (2016)
3. Fehr, E., Camerer, C.: Social neuroeconomics: the neural circuitry of social preferences. Trends Cogn. Sci. **11**, 419–427 (2007)
4. Glimcher, P.W., Rustichini, A.: Neuroeconomics: the consilience of brain and decision. Science **5695**, 447–452 (2004)

5. Goucher-Lambert, K., Moss, J., Cagan, J.: Inside the mind: using neuroimaging to understand moral product preference judgments involving sustainability. J. Mech. Des. **139**, 41–103 (2017)
6. Guixeres, J., et al.: Consumer Neuroscience-based metrics predict recall, liking and viewing rates in online advertising. Front. Psychol. **8** (2017). https://doi.org/10.3389/fpsyg.2017. 01808
7. Hunt, L.T., Dolan, R.J., Behrens, T.E.: Hierarchical competitions subserving multi-attribute choice. Nat. Neurosci. **17**, 1613–1622 (2014)
8. Keeney, R.L., Raiffa, H.: Decisions with Multiple Objectives - Preferences, and Value Tradeoffs. Wiley, New York (1976)
9. Khushaba, R.N.: Consumer neuroscience: assessing the brain response to marketing stimuli using electroencephalogram (EEG) and eye tracking. Expert Syst. Appl. **40**, 3803–3812 (2013)
10. Kothe, C.A., Makeig, S.: Estimation of task workload from EEG data: new and current tools and perspectives. In: Annual International Conference of the IEEE Engineering in Medicine and Biology Society (2011)
11. Laeng, B., Sirois, S., Gredebäck, G.: Pupillometry: a window to the preconscious? Perspect. Psychol. Sci. **7**, 18–27 (2012)
12. Mohr, P.N.C., Biele, G., Heekeren, H.: Neural processing of risk. J. Neurosci. **30**, 6613–6619 (2010)
13. Morin, C.: Neuromarketing: the new science of consumer behavior. Society **48**, 131–135 (2011)
14. Porter, G., Troscianko, T., Gilchrist, I.D.: Effort during visual search and counting: insights from pupillometry. Q. J. Exp. Psychol. **60**, 211–229 (2007)
15. Rangel, A., Camerer, C., Montague, P.R.: A framework for studying the neurobiology of value-based decision making. Nat. Rev. Neurosci. **9**, 545–556 (2008)
16. Riedl, R., Davis, F.D., Hevne, R., Alan, R.: Towards a NeuroIS research methodology: intensifying the discussion on methods, tools, and measurement. J. Assoc. Inf. Syst. **15**, I (2014)
17. Sanfey, A.G., Rilling, J.K., Aronson, J.A., Nystrom, L.E., Cohen, J.P.: The neural basis of economic decision-making in the ultimatum game. Science **5626**, 1755–1758 (2003)
18. Sharma, N., Gedeon, T.: Objective measures, sensors and computational techniques for stress recognition and classification: a survey. Comput. Methods Programs Biomed. **108**, 1287–1301 (2012)
19. Slanzi, G., Balazs, J., Velásquez, J.D.: Predicting Web user click intention using pupil dilation and electroencephalogram analysis. In: IEEE/WIC/ACM International Conference on Web Intelligence (WI). IEEE (2016)
20. Smith, D.V., Huettel, S.A.: Decision neuroscience: neuroeconomics. Wiley Interdiscip. Rev. Cogn. Sci. **1**, 854–871 (2010)
21. Sylcott, B., Cagan, J., Tabibnia, G.: Understanding consumer tradeoffs between form and function through metaconjoint and cognitive neuroscience analyses. J. Mech. Des. **135** (2013). https://doi.org/10.1115/1.4024975
22. Weber, M., Borcherding, K.: Behavioral influences on weight judgments in multi-attribute decision making. Eur. J. Oper. Res. **67**, 1–12 (1993)
23. Zhao, Y.L., Siau, K.: Cognitive neuroscience in information systems research. J. Database Manag. **27**, 58–73 (2016)

Impact of Negotiators' Predispositions on Their Efforts and Outcomes in Bilateral Online Negotiations

Bo Yu[1(✉)] and Gregory E. Kersten[2]

[1] Rowe School of Business, Dalhousie University, Halifax, Canada
bo.yu@dal.ca
[2] J. Molson School of Business, Concordia University, Montreal, Canada
gregory.kersten@concordia.ca

Abstract. This study uses the Thomas-Kilmann Instrument (TKI) to analyze the negotiators' predispositions in handling conflicts in online negotiations. It explores the impacts of the individual predispositions on the negotiation processes and outcomes. The results show that TKI scores are significantly related to both the efforts that the negotiators put in their negotiation activities and the achieved agreements. The results also show that the various compositions of individual predispositions in dyadic negotiations can lead to different results.

Keywords: Bilateral negotiation · Online negotiation experiments
Individual predispositions · Thomas-Kilmann Instrument

1 Introduction

Negotiation is a mechanism frequently used to resolve conflicts or solve problems involving two or more individuals or organizations. During their negotiations negotiators need to evaluate offers and arguments they receive from their counterparts and decide on their own offers and arguments. Individual characteristics influence the negotiation process and its outcomes. However, empirical studies differ in their assessment of the impact of individual characteristics on negotiations. Potential reasons include the negotiators' ability to adapt to different contexts, problems, and counterparts, individual characteristics distorted by situational factors, and the confounding effect of the other party [1–3].

It is a challenge to decide on an effective way to group negotiators into specific categories of characteristics in order to obtain a large enough sample for analysis. In the last decade, the InterNeg Research Centre conducted online experiments for both training and research purposes [4]. More than 1000 dyadic negotiations with anonymous partners have been conducted. To capture the participants' predispositions regarding five conflict-handling approaches, prior to the negotiations they were asked to answer Thomas-Kilmann Instrument (TKI) questions [5]. Based on the data collected from the experiments, the current study investigates the influence of individual

© Springer International Publishing AG, part of Springer Nature 2018
Y. Chen et al. (Eds.): GDN 2018, LNBIP 315, pp. 70–81, 2018.
https://doi.org/10.1007/978-3-319-92874-6_6

predispositions on negotiation by exploring the impact of TKI scores on negotiators' efforts during the negotiations and negotiation outcome. The results show that TKI can be used to distinguish individual negotiators in terms of their general predispositions to conflict resolution.

In most cases the negotiators have strong and medium predispositions to two or three approaches. This allows them to select an approach that they consider the most fitting a particular situation as well as to change the approach during the negotiations [6, 7]. The negotiators who face exactly the same type of conflict and who are placed in the same situational context should employ their strongest and the best-fitting predispositions. The question asked here is as follows: *Do the negotiation predispositions, subject to the perturbations introduced by the anonymous counterparts, influence the negotiators' aspirations and their behavior?*

The analysis of the negotiation data shows that the predispositions indeed influence the negotiators' aspiration levels and the negotiation process and its outcomes.

2 Dual Concern Model and Thomas-Kilmann Instrument

Blake and Mouton [8] proposed the "managerial grid", a model to assess managers' conflict caused by their concern for people and concern for results. Managerial grid offers a perspective on social value orientation that is particularly suitable in studies of, and approaches to, negotiations. To stress its applicability in negotiations, it was renamed as a dual concern model. It has been used in negotiation research and verified in numerous studies [e.g., 9–11]. The adapted model uses the strength of the negotiator's *concern for self* and *concern for others* (counterpart) to determine the negotiation approach predisposition. These two concerns are used to specify the following five predispositions of the negotiators: (1) *avoiding* conflict and disengaging with the counterpart; (2) *accommodating* requests of the counterpart; (3) *competing* with the counterpart to achieve as much as possible; (4) *collaborating* to achieve a solution that satisfies both parties; and (5) *compromising* which involves making and demanding concessions to achieve a solution that both sides can accept [7].

Several research instruments to measure negotiators' predispositions towards the five conflict-handling modes were developed [12]. One of the most widely used instruments to measure the propensity for negotiation approach is TKI (also called MODE) formulated by Thomas and Kilmann [13]. TKI uses a variant of the dual concern model with the dimensions describing assertiveness (effort to satisfy own concerns) and cooperativeness (effort to satisfy the counterpart's concern). It is a forced-choice instrument designed to create an individual profile which is a vector of five values (from 0 to 12) associated with each approach [14]. TKI has been commercialized and used to help individuals understand the impacts of different conflict-handling modes in various settings [15].

3   Inspire Bilateral Online Negotiations

Inspire is an e-negotiation system supporting bi-lateral multi-issue negotiations with enhanced negotiation analytic methods, communication, and dynamic user-controlled graphical tools [16, 17]. The system has been used in both lab and online negotiation experiments.

3.1   Inspire System and Experiments

In 2009 the GRIN project (global research in Internet negotiation) was initialized by a group of researchers and instructors from multiple universities in 5 different countries [4]. The Inspire system became part of the GRIN's activities. Over the last decade, online negotiations via Inspire have been regularly conducted for students and professionals from different countries. The system has been used to augment and enhance courses. For that purpose, teaching materials, lecture notes, slides and assignments were designed. Participants were asked to consent to the collection of their negotiation transcripts and to fill in pre- and post-negotiation questionnaires. The InterNeg Research Centre did not provide any specific incentive to participants, instead the Centre requested that instructors integrate the Inspire negotiation in their courses and use it and the accompanying report as an assignment.

Most of the Inspire users found the system easy and fun to use. They have enjoyed online discussions with unknown opponents. They were able to use different strategies, learn more about negotiations and negotiation support, and work on their communication and negotiation skills.

To provide participants with real-life-like context and enhance their engagement, several business cases were created. One of the most often used cases that young participants from different countries could relate to was "Yowl-Pop". This case involves a music artist and an entertainment company negotiating a contract. The scenario involves four issues, each of them has several options (http://invite.concordia.ca/cases/inspireYowlPop.html). An agreement can be made when the two parties agree upon a contract that contains one option for each issue.

The negotiation process of each Inspire negotiation is divided into three phases: negotiation preparation, negotiation, and post-settlement. Participants were asked to fill in TKI questionnaire, during the negotiation preparation phase. Then, they read the materials related to the case. According to the information contained in the business scenario, they specified their preferences with the issues and options. Before the participants started their negotiations, they were asked to specify the best contract that they may achieve and the worst-but-acceptable contract. In the negotiation and post-settlement phase, the elicited preferences were used to provide decision support to the participants. The utility (score) of the expected and achieved agreements were also measured based on the elicited preferences.

3.2   The Dataset

The participants were paired into dyads; as a rule, students from one university represented one side of the case: either the agent of the musician or the manager of the entertainment company. In total, 1994 individual observations were obtained after cleaning the data. All the participants of the current study answered the TKI questions. The reported age of the majority sample (i.e., 80.2%) was between 20 and 30. The data were collected from eleven online experiments conducted between 2010 and 2016.

The composition of the data in chronological order is reported in Table 1. The dataset of each online experiment is further decomposed according to gender. Overall, the number of female participants is slightly higher than that of male participants.

Table 1. Participants in the Inspire negotiations

Experiment (year/month)	Age (20–30)	Gender			Total
		Female	Male	Missing	
2010/12	195	116	111	8	235
2011/05	122	58	74	75	207
2011/10	138	67	88	0	155
2012/04	107	66	45	0	111
2013/04	219	109	121	0	230
2013/11	87	70	73	0	143
2014/04	253	177	136	0	313
2014/11	53	34	33	0	67
2015/04	233	164	117	0	281
2015/11	31	28	32	0	60
2016/04	161	115	77	0	192
Total	1599	1004	907	83	1994

4 Results

The TKI has been used for over forty years as an instrument to assess general strength of individual predispositions to conflict situations [5]. Inspire users were asked to fill in TKI questionnaire prior their negotiation preparation activities.

4.1 Comparison of CPP and Inspire TKI Results

The CPP Inc. (https://www.cpp.com) developed a report with a normative sample that can be used to guide applications of the TKI instrument and interpretation of its results [18]. The normative sample comprises 8,000 American respondents. The selection of the sample is balanced between males and females. The selection also represents respondents' different levels in organizations, ethnicities, regions of the United States, and so on.

The CPP's normative sample is restricted to the United States. In contrast, the majority of respondents in our dataset are from outside of North America. The comparison of the Inspire sample and the CPP sample was conducted to check for both the consistency and the differences in our dataset and the CPP normative sample. The comparison of TKI raw scores on three level of percentiles is reported in Table 2.

Table 2. TKI raw score comparison of Inspire and CPP samples

Range	Competing		Collaborating		Compromising		Avoiding		Accommodating	
	Inspire	CPP	Inspire	CPP	Inspire	CPP	Inspire	CPP	Inspire	CPP
Top 25%	7–12	7–12	**6–12**	9–12	10–12	10–12	8–12	8–12	7–12	7–12
Middle 50%	3–6	3–6	**4–5**	5–8	7–9	6–9	5–7	5–7	3–6	4–6
Bottom 25%	0–2	0–2	**0–3**	0–4	0–6	0–5	0–4	0–4	0–2	0–3

TKI adopts a forced-choice approach, in which respondents have to make choices between 30 pairs of statements [13]. The choices of the respondents are counted to form the raw TKI scores. The TKI percentile scores are obtained by rescaling the raw score within a sample.

Table 2 shows that the raw sores in competing, compromising, accommodating, and avoiding modes at the top 25% are the same as those of CPP. The scores between 0% and 25%, and between 25% and 75% percentile are close to those of CPP normative sample. The only difference is the collaborating predisposition. Strong collaborating predisposition (i.e. 75%–100%) has a wider score range for the Inspire negotiators than for the CPP sample, i.e., 6–12 vs. 9–12. In contrast, weak collaborating predisposition has narrower score range for Inspire sample than for the CPP sample, i.e., 0–3 vs. 0–4. The score range for medium collaborative predisposition is much narrower for Inspire data than for CPP data, 4–5 vs. 5–8. The comparison results indicate fairly good reliability and validity of TKI in the current research setting.

Possible factors behind this difference include culture, age, occupation, and place of residence. The majority of respondents in our dataset are younger than those in CPP normative sample. Most of them are students from different global regions rather than from the US only. Unfortunately, it is not possible to explore which is the key factor that causes the differences.

4.2   The Effects of TKI on Negotiation Effort and Aspiration

The correlations of TKI scores with a set of measures are examined here in order to determine the effect of TKI scores on the negotiators' aspirations before the negotiations and their effort during the negotiations.

We selected the following four variables to represent the effort: the number of offers, the number of messages, negotiation time, and the length of messages (i.e., the no. of characters). The correlations are given in Table 3.

The negative and significant correlation between the competing score and the number of messages indicates that negotiators with strong competing predisposition tend to send fewer messages. This result also suggests that negotiators with strong competing predisposition may spend less effort persuading their counterparts, since persuasion can only take place in messages in their negotiations. In contrast, negotiators with strong collaborating predisposition tend to send more and longer messages, since the collaborating score significantly correlates with both the number of messages

Table 3. The correlation of TKI scores with process and aspiration

TKI Mode	Effort				Agreement (score)	
	Number of C offers	No. of messages	Nego. time (hours)	Length of messages	Expected best	Worst acceptable
Competing	.017	−.050*	.027	.010	−.002	.021
Collaborating	−.001	.063**	.014	.048*	.030	.041
Compromising	−.081**	.036	.020	.038	.043	.025
Avoiding	.048*	−.027	−.052*	−.043	.016	.005
Accommodating	.012	.023	−.018	−.043	−.078**	−.066**

*significant at the 0.05 level (2-tailed); **significant at the 0.01 level (2-tailed).

and the length of messages. This result indicates that these negotiators put more effort in persuading their counterparts.

The compromising score significantly and negatively correlates with the number of offers, which suggests that the negotiators with strong compromising predisposition send fewer offers. This result also suggests that these negotiators were more likely to wait for their counterparts to propose offers and then they could consider whether the offers were acceptable.

The avoiding score correlates positively with the number of offers and negatively with negotiation time. This result suggests that the stronger the avoiding predisposition of a negotiator, the more offers will be sent but less time will be spent. These negotiators tend to interact less with their counterparts, while they tried more offers that were more substantive to potential agreement.

The correlations between TKI scores and the utility value of the expected best contract and the utility value of the worst contract that negotiators expected were tested. These two values reflect the negotiators' aspiration levels regarding their negotiation. The correlations show that the accommodating score negatively correlates with the expected agreement score and the score of the worst acceptable agreement.

4.3   Relationship Between Negotiators' Predispositions and Agreements

Both the raw scores and percentile scores of the five predispositions are not independently measured in TKI. Thereby, these scores cannot be used as regular variables. The relative strength of the five predispositions determines the general characteristic of a person in terms of their approach to conflicts.

Given the specifics of TKI, the current study uses its scores as dependent variables when examining their potential influence on agreement. A Kruskal-Wallis test was conducted to compare the TKI percentiles of the five scores between two groups: (1) without-agreement group (i.e., 289 members did not reach agreement) and (2) with-agreement group (i.e., 1705 members reached an agreement). The significant difference between the two groups indicate the potential influence of the five predispositions. The test results are presented in Table 4.

The results show that the two groups significantly, albeit weakly, differ in terms of the compromising score at 10% level. There is no significant difference between scores

Table 4. Comparison of TKI scores between with- and without-agreement groups

TKI Mode	Agreement	Mean rank	Significance
Competing	No	1043.14	.143
	Yes	989.76	
Collaborating	No	955.27	.172
	Yes	1004.66	
Compromising	No	938.72	**.058**
	Yes	1007.46	
Avoiding	No	981.48	.606
	Yes	1000.22	
Accommodating	No	1031.98	.267
	Yes	991.66	

for the other predispositions. This result indicates that the stronger the negotiators' compromising predisposition the more likely they are to achieve an agreement.

The influence of TKI score is further examined by checking correlations with the achieved agreement utility for the with-agreement group. The correlations are reported in Table 5. The competing score significantly and positively correlates with agreement utility, which suggests that the stronger the competing predisposition, the greater the utility value of the agreement. On the other hand, the accommodating score significantly but negatively correlates with agreement utility. This suggests that the more accommodating the negotiators are, the lower utility value they achieved in their agreements. Since, most, if not all, agreements require compromise this result is consistent with our expectations.

Table 5. The correlation of TKI scores with agreement utility

TKI mode	Agreement utility
Competing	.062*
Collaborating	.011
Compromising	.003
Avoiding	.007
Accommodating	−.091*

*Correlation is significant at the 0.05 level (2-tailed).

4.4 Tests Aligned with the Triangle Hypothesis

The adaptation of the collaborating negotiators when they negotiate with competitive counterparts, was first mentioned by [19] and confirmed in experiments conducted by Kelley and Stahelski [2], Weingart et al. [20], and others. The triangle hypotheses posit that collaborating negotiators view the negotiation world as comprising of both collaborating and competing negotiators, while competing negotiators see only competing negotiators. This is because collaborating negotiators modify their behavior; when their counterparts compete; the cooperative negotiators will adapt and compete as well.

Competing negotiators, however, do not adapt their behavior; they compete with both collaborating and competing counterparts.

The competing and collaborating TKI scores indicate personal predispositions to be collaborating or competing when individuals handle conflicts. Several tests of competing and collaborating TKI scores aligned with the triangle hypotheses were conducted. The tests were carried out by coding negotiation instances based on the TKI scores of the dual parties. Only the dyads where both parties had TKI scores were selected. TKI scores were firstly used to code individual negotiators' profiles by following the suggestion of Shell [7]. The 75% percentile was used as the cut-off point. TKI scores that are above 75% were coded as indicators that the individual would behave strongly in the respective modes. For instance, if an individual has the following set of TKI scores: competing – 8, collaborating – 5, compromising – 7, accommodating – 6, and avoiding – 4, then this set of scores was first converted to percentiles, i.e., competing – 82%, collaborating – 45%, compromising – 24%, accommodating – 47%, and avoiding – 36%. This individual was then profiled as being strong in competing because only the competing percentile is above 75%.

It is possible that some individuals are strong in both competing and collaborating modes. These negotiators' profiles were temporarily coded as a special case, in which the negotiators' profiles depend on their counterparts. If their counterparts were strong in the competing mode, the negotiators were profiled as competing. If their counterparts were strong in the collaborating mode, the negotiators were profiled as collaborating as well. This coding rule aligns with the propositions of triangle hypotheses.

The coded individual negotiators' profiles were used to further code negotiation instances. For instance, a negotiation instance will be coded as "collaborating-collaborating" if both parties are strong in collaborating mode. Competing-competing indicates that both parties are strong in competing mode. Competing-collaborating suggests that one party is strong in competing mode, while the other party is strong in collaborating mode. The instances were coded as "other" when any party was not profiled as being strong in either collaborating or competing mode. The coded instances were grouped giving their profiles. The between-group differences in terms of agreement were then tested and the results are presented in the Table 6.

Table 6. Instance profiles and agreements

Dyads	Agreement			Sum
	No	Yes	Rate	
Collaborating-collaborating	7	57	89.1%	64
Competing-collaborating	13	50	79.4%	63
Competing-competing	8	41	83.7%	49
Other	112	703	86.3%	815
Total	140	851	85.9%	991

The collaborating-collaborating group has the highest agreement rate, i.e., 89.1%, while the competing-collaborating group has the lowest agreement rate, i.e., 79.4%. A Chi-square test was conducted with the cross-tab approach to examine whether the agreement rate differs between groups. No significant effect was found ($p = 0.387$).

The "other" group was then filtered out after the test. A non-parametric median test was then conducted to examine the difference in terms of joint performance of agreements between groups (i.e., the three groups were profiled in terms of being collaborating and competing). Two measures were adopted in this test, including the number of dominating packages and the joint utility of agreement. The joint utility of agreement was calculated as the product of the two parties' individual utility in each negotiation (Table 7).

Table 7. Instance profiles and joint utility

Joint utility	Dyads		
	Collaborating-collaborating	Competing-collaborating	Competing-competing
>Median	20	30	24
<=Median	37	20	17
Ratio	0.54	1.5	1.41

No significant effect was found in terms of the number of dominating packages. The results show that the collaborating-collaborating group has a significant number of instances with their joint utilities below median (p = 0.016). This finding contradicts the expectation that collaborating-collaborating dyads should achieve better performance. The results are visualized with a box-plot shown in Fig. 1. The box-plot shows that the collaborating-collaborating group has no more observations of high joint utilities than the other two groups.

These results partially confirm the results of earlier studies in which participants negotiated face-to-face [10, 20, 21].

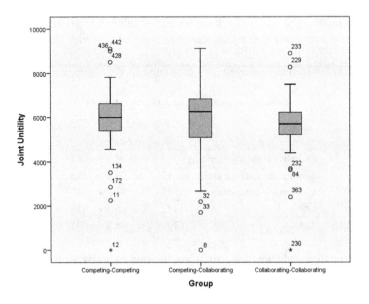

Fig. 1. Plot of instance profiles and instance joint utility

5 Conclusions

This study explores the effects of individual predispositions of handling conflict in negotiations by using TKI. It demonstrates that individual predispositions to resolution of conflict influence negotiations in many ways.

The obtained results indicate that individual predispositions influence the negotiation process. The higher the negotiators' competing scores, the fewer massages they sent to their counterparts. In contrast, the higher the negotiators' collaborating scores, the more and longer messages they sent. This suggests that competitive negotiators are less interested in establishing rapport with their counterparts and educating them. Instead, they are focused on achieving high substantive outcomes.

Negotiators with higher avoiding scores were found to send more offers, while achieving the compromise in shorter time than other negotiators. We also found that compromising scores have negative impact on the number of offers. This suggests that strongly compromising negotiators are less interested in the offer exchange process.

We found that individual predispositions can influence whether an agreement will be reached. Negotiators who reached an agreement had stronger compromising predisposition than those who failed to obtain an agreement. Understandably, negotiators who had stronger accommodating predisposition achieved lower agreement utility. Negotiators who had strong collaborating predisposition did not outperform others in terms of either agreement rate or the utility values of achieved agreements. These negotiators put more effort into their negotiations but their efforts did not produce better results than the results achieved by other participants. Negotiators with stronger competing predisposition achieved higher agreement utility. During the process, they sent fewer messages; this could have helped them to focus on extracting value from their counterparts.

The agreement rate and the performance of negotiation dyads profiled with the combination of strong-to-medium competing and collaborating predispositions of the paired negotiators were also examined. It was found that negotiation instances with both parties having high collaborating scores did worse as compared with the instances with either both parties having high competing scores or the dyads composed of one party with high competing score and the other party with high collaborating score. These findings confirm results of earlier experiments that collaborative dyads more often accept inefficient agreements than competitive dyads [22, 23].

In summary, this study confirms the usefulness of the Thomas-Kilmann Instrument which can be effectively used to characterize individual predispositions. While the predispositions are general in the sense that people behave differently in different situations (e.g., they may compete in one negotiation and collaborate in another), we have shown that the strength of the predispositions affect the participants who face an identical negotiation problem. The impact of the predispositions on the expectations, efforts and agreements may be used in teaching. It may also be used in practice; negotiators may be able to assess their counterparts based on the latter focus on offers and/or argumentation.

A limitation of the current study is that most of our participants were young students. Therefore, their negotiation, judgement, and decision-making behaviors represents closely the young population group. However, the findings of the current study are still applicable to other population groups, if individual predispositions of handling conflicts indeed have impacts on negotiators behaviors. The predispositions are often stable and evolve slowly over time. Future research may reveal more insights into the influence of individual predispositions on negotiation processes and outcomes. The effort and benefit analysis introduced here may be further elaborated. Negotiators may wish to obtain greater benefits with less effort. In the current study, multiple TKI scores (e.g., competing and collaborating) have significant impact on negotiation process and outcome. The application of the enhanced effort and benefit analysis may yield interesting results.

References

1. Butt, A.N., Choi, J.N., Jaeger, A.M.: The effects of self emotion, counterpart emotion, and counterpart behavior on negotiator behavior: a comparison of individual level and dyad level dynamics. J. Organ. Behav. **26**(6), 681–704 (2005)
2. Kelley, A.H., Stahelski, A.J.: Social interaction bases of cooperators' and competitors' beliefs about others. J. Pers. Soc. Psychol. **16**(1), 66–91 (1970)
3. Bazerman, M.H., et al.: Negotiation. Annu. Rev. Psychol. **51**, 279–314 (2000)
4. Paradis, N., et al.: E-negotiations via Inspire 2.0: the system, users, management and projects. In: 2010 Proceedings of Group Decision and Negotiations (2010)
5. Thomas, K.W., Kilmann, R.H.: Thomas-Kimann Conflict Mode Instrument. Xicom, Tuxedo (1974)
6. Greenhalgh, L., Chapman, D.I.: Joint decision making: the inseparability of relationship and negotiation. In: Kramer, R.M., Messick, D.M. (eds.) Negotiation as a Social Process, pp. 166–185. Sage, Thousands Oaks (1995)
7. Shell, G.R.: Teaching ideas: bargaining styles and negotiation: the Thomas-Kilmann conflict mode instrument in negotiation training. Negot. J. **17**(2), 155–174 (2001)
8. Blake, R.R., Mouton, J.S.: The Managerial Grid. Gulf, Houston (1964)
9. Dévényi, M.: The role of integrative strategies and tactics in HR negotiations. Strateg. Manag. **21**(2), 32–36 (2016)
10. Rhoades, J.A., Carnevale, P.J.: The behavioral context of strategic choice in negotiation: a test of the dual concern model. J. Appl. Soc. Psychol. **29**(9), 1777–1802 (1999)
11. Sorenson, R.L., Morse, E.A., Savage, G.T.: A test of the motivations underlying choice of conflict strategies in the dual-concern model. Int. J. Confl. Manag. **10**(1), 24–42 (1999)
12. Rahim, A.M., Magner, N.R.: Confirmatory factor analysis of the styles of handling interpersonal conflict: first-order factor model and its invariance across groups. J. Appl. Psychol. **80**(1), 122–132 (1995)
13. Kilmann, R.H., Thomas, K.W.: Developing a forced-choice measure of conflict-handling behavior: the "MODE" instrument. Educ. Psychol. Measur. **37**(2), 309–325 (1977)
14. Meyer, C.J., et al.: Scissors cut paper: purposive and contingent strategies in a conflict situation. Int. J. Confl. Manag. **23**(4), 344–361 (2012)
15. Schaubhut, N.A.: Technical Brief for the Thomas-Kilmann Conflict Mode Instrument: Description of the Updated Normative Sample and Implications for Use. CPP Inc. (2007)
16. Kim, J.B., et al.: E-negotiation system development: using negotiation protocols to manage software components. Group Decis. Negot. **16**(4), 321–334 (2007)

17. Kersten, G.E., Noronha, S.J.: WWW-based negotiation support: design, implementation, and use. Decis. Support Syst. **25**, 135–154 (1999)
18. Schaubhut, N.A.: Thomas-Kilmann conflict mode instrument. CPP Research Department (2007)
19. Sermat, V., Gregovich, R.P.: The effect of experimental manipulation on cooperative behavior in a chicken game. Psychon. Sci. **4**(12), 435–436 (1966)
20. Weingart, L.R., et al.: Conflicting social motives in negotiating groups. J. Pers. Soc. Psychol. **93**(6), 994 (2007)
21. Thompson, L.: Negotiation behavior and outcomes: empirical evidence and theoretical issues. Psychol. Bull. **108**(3), 515 (1990)
22. Beersma, B., et al.: Cooperation, competition, and team performance: toward a contingency approach. Acad. Manag. J. **46**(5), 572–590 (2003)
23. Olekalns, M., Smith, P.: Social value orientations and strategy choices in competitive negotiations. Pers. Soc. Psychol. Bull. **25**(6), 657–668 (1999)

Some Methodological Considerations for the Organization and Analysis of Inter- and Intra-cultural Negotiation Experiments

Tomasz Wachowicz[1]([✉]), Gregory E. Kersten[2], and Ewa Roszkowska[3]

[1] University of Economics in Katowice, ul. 1 Maja 50, 40-287 Katowice, Poland
tomasz.wachowicz@ue.katowice.pl
[2] J. Molson School of Business, Concordia University, Montreal, Canada
gregory@jmsb.concordia.ca
[3] Faculty of Economy and Management, University of Bialystok,
Bialystok, Poland
erosz@o2.pl

Abstract. In this paper we analyze some problems related to the design and analysis of the inter- and intra-cultural online negotiation experiments in which university students participate. We discuss factors that may impact the negotiation performance. Apart from national culture, which is an evident factor the impact of which is traditionally measured in cross-country negotiations, we discuss also the potential influences of university or students' individual or group culture. When analyzing the negotiation performance, we focus not only on the bargaining process, but also on the pre-negotiation preparation. The paper provides a statistical analysis of the negotiation experiments organized in Inspire, which – unfortunately – were not designed to study all the issues raised in the paper, yet allow us to capture some of the ideas and notions discussed.

Keywords: National culture · Individual culture · Instructions
Pre-negotiation · Negotiation outcomes · Online negotiation
Negotiation experiments

1 Introduction

Internet and related technologies created new opportunities to study social-psychological and economic interactions among people. These include negotiations, leading to numerous online negotiation experiments, which aim at analyzing the impact of different technologies on the negotiation process and outcomes (e.g. [1–3]); comparing face-to-face and online negotiations (e.g. [4–6]); and analyzing behavioral patterns of the negotiators and their evolution (e.g. [7–9]). The later type of experiments includes inter-cultural and inter-organizational studies in which participants from different cultures are paired up (e.g. [10–12]).

The design and conduct of online negotiation experiments requires paying attention to details. The researchers in both lab and online experiments, need to prepare a case, establish the experiment protocol and design its process, determine and implement the controlled factors, and clearly define factors that they cannot control. In the case of online negotiations, additional challenges arise from the experimenters located in different sites, in the site-specific context (e.g., type of the organization and extent of supervision), and the participant-specific situation (e.g., place and time, time-pressure). The researchers need also to consider the possible differences in the participants' understanding of the case, protocol, instructions, etc.

To illustrate the difficulties, we briefly discuss three cross-cultural studies which report on cultural differences when negotiations are conducted online.

A three-country online negotiation experiment led the authors to conclude that "Our results suggest that computer-mediated negotiations ... are significantly influenced by the culture the negotiator comes from." [11, p. 505]. The authors note that one limitation of their study is that the individualism/collectivism cultural dimension, which was the focus of the study, was not directly measured. They realized that a direct measurement of a cultural dimension may lead to ambiguity or loss of clarity of the cultural dimension, because individual characteristics of participants from different cultures may be more similar than the characteristics of the participants from the same culture.

In another study, members of Amazon's Mechanical Turk were invited to participate in an experiment in which they were asked to assess an opening move made by, they were supposed to believe, their negotiation counterpart (who was, in fact, the experimenter). They were classified on the basis of their nationality (100% U.S.) and ethnicity (80.8% European and 19.2% others) [13]. The authors of this study analyzed the reaction of this group to their counterpart's expression of anger, which the group members could not verify; anger and the counterpart's culture (either East Asian or European) were the control variables. The study's results showed that "cultural background significantly shapes the effects of expressing anger in negotiations. Angry East Asian negotiators being perceived as tougher and more threatening compared with angry European American negotiators" [13, p. 795]. Introducing cultural distinctions among sub-national groups causes that the term 'national culture' becomes ambiguous. Furthermore, the differences in the perception of the same anger "labeled of an East Asian" or "labeled of a European" may be due to the perception of sameness rather than any actual differences.

The last example concerns email negotiations between two groups of students from Hong Kong and the U.S. [12]. The experiment was very carefully prepared; the instructors standardized their first week of classes using the same materials and teaching notes; about 50% of both Hong Kong and U.S. students played the role of the seller and 50% played the role of the buyer. A different case was prepared for each sub-group. The experiment results showed that Hong Kong sellers were more aggressive and negotiated higher prices than their U.S. counterparts, which could be consistent with a cultural reactance effect for the Hong Kong's students, or an alternative explanation could be the possibility that in intercultural negotiations Asians become more individualistic

[12, p. 636]. In order to sharpen the results, two more experiments were conducted: face-to-face and email intracultural negotiations. The results are surprising; although the authors observe that "culture would moderate the effect of communication media on opening offer amount, which in turn would mediate the effect on price per episode." [12, p. 638], the results from the inter- and intra-cultural experiments show (op. cit., Tables 1 and 2) that culture plays no role in the opening offers. Hong Kong students made aggressive opening offers in email negotiations, both inter- and intra-cultural. Their offers in intracultural negotiations were more aggressive than in intercultural negotiations. In face-to-face negotiations, there was no difference between Hong Kong and U.S. students. Given that initial offers form anchors and are strong predictors of the final outcomes, it is not possible to state that culture affects negotiation and its outcomes [14]. The experiments show that the Hong Kong but not the U.S. students are more aggressive when they negotiate via email, but the presented results do not provide any insight into this trait.

The above three examples illustrate some of the difficulties in studying online negotiations from the cultural differences perspective. The goal of this paper is to look into the possible problems and errors that may occur when the experiments' purpose is to determine cultural similarities and differences in analyzing the negotiation processes. In particular, we look into the possible moderation of the impact of culture on the negotiation performance by organizational culture and by the students' perception of the instructions. Although the instructors may use the same instructions and materials, students from different universities and different cultures may understand the instructions differently. Furthermore, when analyzing the negotiation performance, we focus on both negotiation phases: preparation and bargaining. We do not consider the post-negotiation phase, while this phase occurs in real life, it is rare in negotiation experiments. Pre-negotiation, however, is the critical but not well researched phase. Jang, Effenbein and Bottom's [15] meta-analysis led the authors to conclude that while "the field converged upon measuring and modeling the bargaining phase using experimental methods, to the relative exclusion of other phases and methods". We will try to find how students from different cultures and universities perform in pre-negotiation and if the pre-negotiation quality impacts also the negotiation outcomes.

To answer questions on the impact of culture, organization, and instructions in preparation and bargaining we use an existing database from past negotiations conducted via the Inspire system [16]. The experiments were not designed to address these issues; therefore, we cannot provide the answers to all questions. We provide some answers and also mention other issues that need to be considered in online negotiation experiments. Therefore, and also because we were not able to find any relevant studies which reported both preparation and bargaining phases, this work is exploratory.

2 Some Constructs Relevant for Online Negotiation Experiments

2.1 National and Organizational Culture

Culture can be considered at many levels, including national, regional, organizational and team. One may also distinguish professional and generational cultures. These levels are important because they show the complexity in a specification of the cultural traits of a group of people and their classification to the particular cultures.[1]

National culture has been considered as a constraining factor of organizational culture [17] or that the latter 'mirrors' the former [18]. However, Gerhart and Fang's [19] re-analysis of Hofstede's data show that mean cultural differences between countries are small relative to differences between organizations within a country. Studies also show that some organizations' culture resembles the culture of another country more than their own and that globalization has a stronger effect on organizational culture than on national culture [20, 21]. These and similar results suggest the following proposition:

P1: *If experiments have large groups of participants from a few organizations, then greater focus should be placed on the organizational culture than on national culture.*

Lok and Crawford [22] surveyed managers from Australia and Hong Kong to determine the impact of organizational culture, leadership, and demographic variables on job satisfaction and organizational commitment. Statistical analysis of the combined data produced a model in which education, types of organizational culture, and the leadership style factors had significant impact on job satisfaction and commitment. However, national culture had no significant effect. The national culture difference became significant only through interaction with independent variables: innovative and supportive organizational cultures and with age and gender. This shows that national culture may have a moderating effect on organizational culture. This result, results reported by [12] and similar results lead to the following proposition:

P2: *If national culture is studied in cross-cultural experiments, then it should to be considered as a moderating or mediating variable.*

Note that in the three cross-cultural experiments discussed in Sect. 1, organizational culture was not measured. The focus on national as opposed to organizational culture seems to be typical for the cross-cultural negotiation studies, particularly when students are used. Several studies on university organizational culture and culture of their students provide insights into the difficulty in collecting data on university culture from students and including it in the analysis [e.g., 23–25]. These studies measured the impact of the university on student general achievement, bonding, and engagement. As far as we know, culture of university students treated as a group or quasi profession has not been assessed, which may be due to the open and changing university culture, the

---

[1] We are indebted to one of the reviewers who pointed out that organizational culture may have a moderating effect on the national culture.

changing cultural make-up of the student population and their perceptions, and the weak relationship between university as perceived by its employees and by its students. Instead, students' national culture was assumed (e.g. [10–13]). In light of the recent studies on national and organizational cultures, this approach needs to be revisited.

Cultural variables describe social constructs but they are obtained from aggregation of individual-level data. These constructs are used in experiments to indicate the differences between groups of participants. The assumption is that, on average, members of one cultural group can be described by similar values of the cultural variables; another cultural group is described by different values [26]. One problem with this assumption is that these values may differ much more between individuals in one country than between two countries [27, 28]. Another problem occurs when the cultural group comes from a single organization rather than being randomly drawn from the entire population. As we mentioned, organizations have their own culture that can be moderated by, but different from their national culture. Individuals cannot be described by categorical independent cultural variables but by partial and plural variables [29]. National (societal) culture is a latent, hypothetical construct that affects individuals indirectly, through various institutions [30]. This suggests that there are no grounds to assume that mean responses from a group of university students would in any way correspond to the mean values of the students' national culture. Therefore, we formulate the following proposition:

P3: *Students' cultural traits should be assessed independently of their national and university cultures.*

There are several implications of this postulate. One is that culture of every participating student group should be determined separately from other groups. Another implication is that some students from one country may culturally belong to another country. Furthermore, it is also possible that all students from one country do not fit the mean values associated with the culture of this country. It follows from Proposition 3 that rather than considering national cultures in online experiments involving groups from geographically dispersed organizations, cultures of every group need be considered separately. The emphasis is on the distinction between groups, irrespective of their participation in inter/intra-national, rather than differences based on national and/or organizational culture.

2.2 Mono- and Multi-cultural Groups

In many universities international students are a significant percentage of the student population. In countries with high immigration levels, students may reside in that country for a few or several years. In the past, if these students participated in experiments, they would be removed for the purpose of analysis [12].

Comparisons of mixed-heritage and multi-ethnic individuals show cultural and social differences. Mixed-heritage individuals have been found to have greater cognitive flexibility, greater multi-cultural competence and greater cultural empathy [31, 32]. Comparisons of multi-cultural and mono-cultural (homogenous) groups showed host-visitor and in-group-outgroup classification to occur in the former but not in the latter [33]. This suggests that students who interact with foreign students may have greater ability to engage in online negotiations with unknown counterparts and greater capacity to build an understanding of the others' needs. Conversely, students who come from the same culture and have no interaction with foreign or immigrant students may find building rapport and understanding with unknown counterparts difficult. This leads us to make the following proposition:

P4: *Groups of students who belong to the same national culture but are from differently internationalized universities should be considered as separate cultural groups.*

This proposition does not suggest that these groups should not be compared to their national culture but that their cultural traits should initially be considered separately. Following the initial analysis, the (dis)similarity between the group culture and the national culture may be concluded.

2.3 Organization of Multi-site Experiments

When the experiments are organized at a few universities and across student groups that have different teachers, problems of providing the participants with equal information about the experiment arise. Different instructions, handouts, glossaries or slideshows used by the teachers may provide the participants with different knowledge about the experiment, the case and roles, the usage of potential software technologies, etc. As a result, the students' level of preparedness is not equal, they may differently understand the assignment, its goals and the consequentiality of the experiment. They may define their own goals differently and vary in their engagement in the assignment. For that reason the standardization of the teaching materials, assignment descriptions or the whole courses is highly recommended [12]. It is also important to verify, if all the instructions given to the students were properly understood.

The studies on cultural differences in perception have been conducted for decades [34]. Among others, their impacts on visual perception or perception of speech were extensively analyzed. Some recent studies show how culture may influence the thinking styles and negotiator's cognitions [35]. Nisbett and Miyamoto [36], for example, provide arguments that visual perception in Americans is more analytical, while in Asians it is more holistic. In other studies, De Paulo and Friedman [37] showed that a significant proportion of a message's meaning comes from its associated visual and verbal cues, so if the written text lacks such cues, the negotiators have to rely more on logical argumentation and the presentation of facts, which favors more the individualistic rather than collectivistic cultures. This suggests that different culture groups may have different capabilities to understand written and graphical instructions regarding, for instance, the preferences that should be represented by the participants. Consequently, the participants from different culture groups may set different priorities and goals and, finally, they may negotiate different contracts. Hence, we formulate the following proposition:

P5: *The students from different culture groups may vary in processing and understanding the same instructions given to them by the teachers prior to the negotiation experiments.*

2.4   Pre-negotiation Performance

The vast majority of experimental studies focuses on analyzing the effects of culture on the negotiation process and outcomes. In such analyses, the negotiation process is usually understood as the actual negotiation phase, i.e. bargaining. The studies rarely, if ever, touch the elements of the pre-negotiation phase. If they do, they analyze only some of the pre-negotiation opinions formulated in questionnaires administered prior the negotiations [38]. As far as we are aware, there is no research that focuses on analyzing the influence of cultural differences on the pre-negotiation activities, preparation tasks and their impact on the negotiation process and outcomes.

The pre-negotiation is considered to be the fundamental part of the negotiation process [39, 40]. During the pre-negotiation, the parties prepare for the actual negotiation process, define the negotiation problem, set up the priorities, specify their preferences, define aspiration and reservation levels and formulate the negotiation strategy. The scoring system that describes quantitatively the negotiator's preferences is also built, and used to provide the negotiator with the decision support during the bargaining phase and in post-negotiations [41]. Hence, it may impact the results the parties obtain. Pre-negotiation preparation requires processing information, analyzing and planning, therefore, using the same rationale as in Sect. 2.3, we may formulate the following proposition:

P6: *The students from different culture groups may vary in the pre-negotiation performance.*

3   Experiment and Its Participants

To assess the propositions formulated in Sect. 2, we analyzed the dataset of online bilateral negotiation experiments conducted in the Inspire system [16]. Note, that these experiments were not purposely designed to study the issues related to cultural influences in negotiations. Therefore, we were not able to give precise answers to some questions asked. However, the imperative problems we faced confirmed how complicated the analyses of culture can be and how important it is to design such experiments thoroughly and comprehensively.

3.1   Inspire Experiment Organization

In the Inspire experiments two representatives negotiated a contract over four issues. The case description provided them with detailed information regarding the priorities, goals and expectations of their principals that should be followed in the negotiation. This description formed clear and unambiguous instructions that were the same for Inspire experiment participants playing the same negotiation role. They were given in English, which was not the native language for a significant number of participants, who were the students of nine universities from America, Europe and Asia. Apart from the Inspire instructions, no additional unified handouts, lectures nor slideshows were prepared, however, the teachers organized the introductory lectures for the students discussing the details of the assignment.

The experiment had two phases: (1) preparation for the negotiation; and (2) bargaining. Pre-negotiation preparation included learning about the role, the context, and the system. Note that Inspire experiments involve representative negotiations in which the agents negotiate on behalf of their principals. The fact that students play a role (e.g. the procurement managers) makes pre-negotiation preparation even more important. The preference systems defined in a form of quantitative scoring systems should reflect the preferences of the principals adequately so the support offered to the agents helps them to obtain good results for their principals. Therefore, in representative negotiations the quality of the pre-negotiation preparation may be measured by the extent to which an agent's scoring system is concordant with the principal's preferences. This was measured in our experiment by means of the notion of ordinal and cardinal inaccuracy [42].

During the bargaining phase, the negotiators were exchanging offers and messages to reach an agreement. The scoring systems they had built were used by Inspire to evaluate the negotiation offers and visualize the negotiation progress on the negotiation history graph. For more details of Inspire experiments refer to [43, 44].

In our experiments there were 1297 participating students from 9 universities and of 65 different groups. To make sure that the side characteristics did not unintentionally impact the understanding of instructions, as well as engagement and performance of different student groups, we focused on analyzing the results of one party only. After removing incomplete records, we obtained a dataset containing 295 records. The students came from three universities from two European countries. In the first university (Uni1_C1) there were two groups of students, domestic, representing the same national culture G1 and foreign, with students from various national cultures G2. Similarly, in the second university (Uni2_C2) from the second country, different from Uni1_C1, there were two groups of mono-culture domestic students G3 and multi-cultural foreign students G4. Finally, in the third university Uni3_C2 (from the same country as Uni2_C2) there was only one group of domestic students G5, that share the same nationality as group G3. Both mono- and multi-cultural groups within each university Uni1_C1 and Uni2_C2 had the same teachers. Note, that the potential differences in negotiation performance observed among these groups may be influenced in fact by a mix of cultural traits related to national, university and student culture.

3.2   Participants

The basic characteristics of each group within each university are given in Table 1.

Table 1. The characteristics of the participants

Characteristic	Uni1_C1		Uni2_C2		Uni3_C2	Sign.
	G1	G2	G3	G4	G5	(K-W)
Number of students	53	27	150	25	40	
Gender (% of females)	71.7	55.6	65.7	36.0	67.5	**0.041**
English proficiency[#]	6.45	5.81	4.07	5.28	4.05	**<0.001**
Negotiation knowledge[#]	3.21	3.07	2.48	2.12	2.83	**0.002**
Case understanding[#]	5.96	5.41	4.97	5.28	5.23	**<0.001**

[#]7-point Likert scale (1 – low; 7 – high)

The disproportions in group sizes do not allow us to use neither the ANOVA nor *t*-tests to analyze the between-group differences. However, non-parametric equivalent tests (though, of weaker discrimination power) can still be applied. We used mainly Kruskal-Wallis (K-W) and Mann-Whitney (M-W) tests to analyze differences in variable structures and a fraction test (based on Fisher-Snedecor distribution) to the differences in proportions.

As we can see, the groups are not uniform with respect to the characteristics describing demographics, skills and knowledge. It is interesting that the case understanding differs significantly between G1 and G3, G1 and G5. This could prove that the national cultural differences (mediated perhaps by English proficiency) influence understanding the instructions. However, G1 also differs significantly from G4, but G4 does not differ from G3 nor G5. The former may confirm that national culture does not always impact the instruction understanding, while the latter, that university or group culture may also be an influencing factor here.

This observation addresses our propositions P3 to P5 and simultaneously shows that unifying the instructions may not be a sufficient solution to provide all groups with the same knowledge and understanding of the assignment, and, consequently, to assure comparable analytical conditions. It may be that, contrary to what Rosette et al. recommend [12], different instructions and introductory lectures should be designed for various groups depending on their skills, knowledge and cultural traits.

4   Results

To illustrate selected points discussed above we conduct two analyses of the experimental data. First, we focus on investigating the differences in general negotiation performance between different student groups. Then we conduct a more detailed analysis to identify the set of characteristics that significantly influences the negotiation outcomes.

4.1   Inter and Intra Cultural Differences

To verify the differences in the factors describing the pre-negotiation performance, bargaining style and results between subsequent groups of students the cluster analysis was used. Selected characteristics recorded in Inspire database are shown in Table 2.

Table 2. Pre-negotiation quality and performance

Item/Factor	Uni1_C1		Uni2_C2		Uni3_C2	Sign.
	G1	G2	G3	G4	G5	K-W
Pre-negotiation quality						
Scoring system inaccuracy:						
Ordinal	4.23	4.03	2.30	0.44	2.70	**<0.001**
Cardinal	83.81	75.33	57.77	38.84	69.43	**<0.001**

(*continued*)

Table 2. (*continued*)

Item/Factor	Uni1_C1		Uni2_C2		Uni3_C2	Sign.
	G1	G2	G3	G4	G5	K-W
Negotiation process						
No. of offers + messages	3.06	3.59	3.51	4.96	4.08	**0.001**
Negotiation length (days)	2.26	1.81	2.90	3.32	2.80	0.055
Informativeness (message length)	406.00	528.30	323.83	272.56	221.52	**0.003**
Outcomes						
Agent's score	77.11	72.26	81.43	81.48	82.03	0.091
Principal's score	74.85	76.56	80.17	80.76	79.28	**0.018**
|Agent − Principal|	11.36	12.07	5.91	5.28	7.65	**0.006**

When analyzing the differences in pre-negotiation quality, the necessity of simultaneous analysis of national, students, and university culture becomes evident (recall proposition P1 to P3). Looking at the ordinal inaccuracy (describing the extent to which the agent's scoring system reflects an order of principal's preferences correctly) we can find that some groups differ significantly. If our dataset had been limited to records describing homogenous national culture and only groups G1, G3 and G5 were compared, one could have concluded that national culture differentiates the pre-negotiation accuracy (Mann-Whitney tests confirm that accuracy of G1 is significantly worse than of G3 and G5, with $p < 0.001$).

The distinction between G1, G3 and G5 includes an implicit influence of university culture too, therefore one may argue that these differences may result from some specific characteristic of university culture (though they do not play a significant role in the case of the two universities from the same country – Uni2_C2 and Uni3_C2 – where the difference between 2.30 and 2.70 is not significant, $p = 0.659$). Further, someone could argue that in the case of Uni1_C1 its culture (but neither the students' nor the national one) must have played a key role in the pre-negotiation performance, because the results of G1 do not differ significantly from G2 - the multi-cultural group of students at Uni1_C1 ($p = 0.582$). Unfortunately, the example of Uni2_C2 shows that there can be some national or student culture factors that play a significant role, since the group G4 is strongly better than G3 ($p < 0.001$). These differences address to some extent our proposition P6.

Note further, that the pre-negotiation results discussed above do not correspond so evidently to the negation outcomes. This time, neither national nor student culture seem to play simultaneous role in influencing the negotiation performance (measured by the principal's score of agreement rating), the differences between G1 and G2, and between G3 and G4 are insignificant ($p = 0.40$ and $p = 0.497$ respectively). But when a combined influence of university and national culture is analyzed, students from G1 appear to obtain significantly worse results than those from G3 and G4 ($p < 0.004$). Yet, students' culture may still interfere since G2 does not differ significantly from G3 ($p = 0.091$).

It is also worth noting, that we were unable to strongly prove P4 in our cluster analysis as G3 did not differ significantly from G5 in anything but the level of understanding of the bargaining process ($p = 0.005$).

4.2 Factors Influencing the Negotiation Outcomes

An advanced structural analysis seemed the best tool to explore the nuances in the relationship between the negotiation outcomes and independent variables describing the groups' culture, their skills and knowledge, instructions and pre-negotiation performance. Structural modelling can be very informative especially when linear dependencies occur among the factors in the structure. However, a series of single-criterion curve estimation analyses that we conducted to explore the data and identify the best fitted relation models showed that most of interactions are best represented by quadratic or cubic functional relationships rather than linear ones. Therefore, we decided to implement a multivariate polynomial regression analysis (MPR) to explore the structure of all aforementioned relations. MPR has been used earlier to specify and test discrepancy relationships with a high degree of precision, e.g. in cross-cultural studies of, for example, physical attractiveness [45]; and the expats adjustment [46].

Using the significant variables from our single criterion dependence analyses, we estimated the series of multivariate polynomial models looking for the best fit. The negotiation knowledge (polynomial relationship) and cardinal accuracy (cubic relationship) appeared to be the only significant factors, apart from culture. The final model (with $R^2 = 0.311$) operates also with the set of dummy variables indicating different groups across national, university and individual culture. Group G1 was considered as a control one, hence binary switches x_G2 to x_G5 represented subsequent groups G2 to G5.

Table 3 shows that out of many factors describing the skills of groups, only one variable appears to be significant in predicting the negotiation outcome, i.e., the negotiation knowledge. No significant influence of instruction understanding is confirmed. Cardinal inaccuracy impacts the final results negatively, as well as some mixtures of university and students' culture. Confidence intervals for x_G2 and x_G4 includes zero, which means that neither G2 nor G4 differ significantly in terms of results from G1. This indicates that the university culture may not influence the results alone but may (or may not) be moderated by the national and students' culture (for example, for G3 vs. G4, but not for G1 and G2). Yet, G3 and G4 differ from G1 significantly. G3 students negotiated higher agreement ratings for their principals (3.1 points on average) than the students from G1 and than G5 students (4.22 points on average), given that students from these groups have the same negotiation knowledge and cardinal accuracy. Note, that these are the groups of the same mono-culture students and some national culture effect may be indicated by the significant impact of x_G2 and x_G4 indicators (not confirming, again, the differences related to P4). However, we cannot say anything about the differences in university cultures of G3 and G5, or on the impact of other variables (note that R^2 is rather low), e.g. the one indicating the fact that for G3 and G5, the instructions could have been given in their native languages.

Table 3. Polynomial regression model for agreement's rating as dependent variable

Model (variables)	Estimate	Std. error	95% confidence interval	
			Lower bound	Upper bound
(negotiation knowledge)[a]	1.193	0.322	0.558	1.827
(negotiation knowledge)[b]	−0.231	0.062	−0.353	−0.109
(cardinal inaccuracy)[b]	−2.192E−6	0.000	−2.655E−6	−1.729E−6
x_G2	1.779	1.773	−1.710	5.268
x_G3	3.105	1.244	0.656	5.554
x_G4	3.441	1.879	−0.258	7.139
x_G5	4.216	1.606	1.055	7.377
Constant	75.079	1.486	72.154	78.005

[a]negotiation knowledge
[b]cardinal accuracy

Problems with unambiguous interpretation of mixes of national and university cultures may suggest the validity of our proposition P3, which recommends analyzing students' groups irrespectively of their national or university cultures but as the ones characterized by the individual cultures of the participants.

5 Conclusions

The purpose of this study was to address and discuss the problems and concerns related to the organization and analysis of international negotiation experiments. We tried to point out some issues that need to be taken into account and measured in such experiments since they may affect the final analysis of the negotiation performance, especially when the participants are students. Among many issues the cultural differences seem to be most evident. However, as shown, the differences attributed to culture may be national, regional, organizational (university culture) or even individual, because they may result from the fact that foreign (Erasmus) students behave in a particular way studying abroad and trying to adopt to foreign university culture and, e.g. aim to obtain better grades than at their domestic universities. Other issues are related to the handouts and the instructions given by the teachers to the participating students (teaching materials, handouts, quizzes, transparencies etc.) that may affect their attitudes and engagement. We also raised an issue of focusing on the pre-negotiation activities as they may mediate in relationships some factors have with the outcomes.

Note that apart from the issues discussed here, there are still many other factors such as students' motivations, conflict attitudes, or thinking styles, which can also affect the negotiation process (including pre-negotiation effects) and outcomes the participants obtain. This is confirmed by the results of our polynomial regression. The model discussed in Sect. 4.2 showed only the nonlinear impact of the negotiation knowledge, cardinal error and variables describing the bundles of cultural influences (mixes of individual, university and national cultures) on the negotiation outcome.

However, its fit measure ($R^2 = 0.311$) confirms that there is a significant amount of outcome variance (69%) that could be explained by other variables like those mentioned above.

We also showed that the comparison of the results from different perspectives is required to get some deeper insight into the potential relationships among factors. The cluster analysis focused on distinguishing the differences for various student groups (though, due to the fact that the culture was not directly measured in Inspire experiments, we could not unambiguously state what cultural effect can be assigned to each group) and confirmed that groups vary in terms of skills (Table 1), pre-negotiation quality, negotiation process and outcomes (Table 2). Yet, this did not show the relationships among those variables. We should take into account the fact that influence of some of factors may be quashed by the influence of others and that the relationships do not need to be linear. This was shown with the polynomial model in which the mediation of pre-negotiation accuracy could also be tested. The difference in the negotiation outcomes between G1 and G2 groups which were indicated as significant with Mann-Whitney test, seems to have no significant influence on outcomes, when the additional impact of cardinal inaccuracy and knowledge was included in the model (see x_G2 confidence interval).

Note, that other analytical approaches may also be used in similar analyses of negotiation experiments. Using factor analysis may allow to build aggregates of factors and simplify their structure so the use of advanced path or structural models could be easier. The strength of moderation effects of cultural variables could also be investigated in the current polynomial model using some alternative approaches based on a series of within-group model estimations and comparisons of differences in their structures. Our future work will focus on designing and analyzing new experiments that will include new profiling and cultural variables and implement comprehensive analytical mechanisms as the ones mentioned above.

Acknowledgements. We wish to thank two reviewers who provided valuable comments and suggestions to an earlier version of the paper. This research was supported by the grant from Polish National Science Centre (2016/21/B/HS4/01583) and from the Natural Sciences and Engineering Canada.

References

1. Johnson, N.A., Cooper, R.B.: Media, affect, concession, and agreement in negotiation: IM versus telephone. Decis. Support Syst. **46**(3), 673–684 (2009)
2. Ow, T.T., O'Neill, B.S., Naquin, C.E.: Computer-aided tools in negotiation: negotiable issues, counterfactual thinking, and satisfaction. J. Org. Comput. Electron. Commer. **24**(4), 297–311 (2014)
3. Purdy, J.M., Neye, P.: The impact of communication media on negotiation outcomes. Int. J. Conflict Manage. **11**(2), 162–187 (2000)
4. Chen, I.-S., Tseng, F.-T.: The relevance of communication media in conflict contexts and their effectiveness: a negotiation experiment. Comp. Hum. Behav. **59**, 134–141 (2016)
5. Citera, M., Beauregard, R., Mitsuya, T.: An experimental study of credibility in e-negotiations. Psychol. Mark. **22**(2), 163–179 (2005)

6. Jonassen, D.H., Kwon, H.: Communication patterns in computer mediated versus face-to-face group problem solving. Educ. Tech. Res. Dev. **49**(1), 35–51 (2001)
7. Pesendorfer, E., Graf, A., Koeszegi, S.: Relationship in electronic negotiations: tracking behavior over time. Zeitschrift fur Betriebswirtschaft **77**(12), 1315–1338 (2007)
8. Swaab, R., Postmes, T., Neijens, P.: Negotiation support systems: communication and information as antecedents of negotiation settlement. Int. Negot. **9**(1), 59–78 (2004)
9. Vetschera, R.: Preference structures and negotiator behavior in electronic negotiations. Decis. Support Syst. **44**(1), 135–146 (2007)
10. Gelfand, M.J., Christakopoulou, S.: Culture and negotiator cognition: judgment accuracy and negotiation processes in individualistic and collectivistic cultures. Organ. Behav. Hum. Decis. Process. **79**(3), 248–269 (1999)
11. Graf, A., Koeszegi, S.T., Pesendorfer, E.M.: Intercultural negotiations in interfirm relationships: an international study of electronic negotiation behavior. J. Manag. Psychol. **25**(5), 495–512 (2010)
12. Rosette, A.S., et al.: When cultures clash electronically: the impact of email and social norms on negotiation behavior and outcomes. J. Cross Cult. Psych. **43**(4), 628–643 (2012)
13. Adam, H., Shirako, A.: Not all anger is created equal: the impact of the expresser's culture on the social effects of anger in negotiations. J. App. Psych. **98**(5), 785 (2013)
14. Galinsky, A.D., Mussweiler, T.: First offers as anchors: the role of perspective-taking and negotiator focus. J. Pers. Soc. Psychol. **81**(4), 657–669 (2001)
15. Jang, D., Elfenbein, H.A., Bottom, W.P.: More than a phase: planning, bargaining, and implementation in theories of negotiation (2017)
16. Kersten, G.E., Noronha, S.J.: WWW-based negotiation support: design, implementation, and use. Decis. Support Syst. **25**(2), 135–154 (1999)
17. Hofstede, G.: Cultures and Organizations: Software of the Mind. McGraw-Hill, New York (1997)
18. Javidan, M., et al.: Conceptualizing and measuring cultures and their consequences: a comparative review of GLOBE's and Hofstede's approaches. J. Int. Bus. Stud. **37**(6), 897–914 (2006)
19. Gerhart, B., Fang, M.: National culture and human resource management: assumptions and evidence. Int. I. Hum. Res. Manage. **16**(6), 971–986 (2005)
20. Festing, M.: Strategic human resource management in Germany: evidence of convergence to the US model, the European model, or a distinctive national model? Acad. Manage. Perspect. **26**(2), 37–54 (2012)
21. Ryan, A.M., et al.: Going global: cultural values and perceptions of selection procedures. Appl. Psychol. **58**(4), 520–556 (2009)
22. Lok, P., Crawford, J.: The effect of organisational culture and leadership style on job satisfaction and organisational commitment: a cross-national comparison. J. Manage. Dev. **23**(4), 321–338 (2004)
23. Libbey, H.P.: Measuring student relationships to school: attachment, bonding, connectedness, and engagement. J. Sch. Health **74**(7), 274–283 (2004)
24. MacNeil, A.J., Prater, D.L., Busch, S.: The effects of school culture and climate on student achievement. Int. J. Leadersh. Educ. **12**(1), 73–84 (2009)
25. Sporn, B.: Managing university culture: an analysis of the relationship between institutional culture and management approaches. High. Educ. **32**(1), 41–61 (1996)
26. Leung, K.: Cross-cultural differences: individual-level vs. culture-level analysis. Int. J. Psychol. **24**(6), 703–719 (1989)
27. Desmet, K., Ortuño-Ortín, I., Wacziarg, R.: Culture, ethnicity, and diversity. Am. Econ. Rev. **107**(9), 2479–2513 (2017)

28. Fischer, R., Schwartz, S.: Whence differences in value priorities? Individual, cultural, or artifactual sources. J. Cross Cult. Psychol. **42**(7), 1127–1144 (2011)
29. Morris, M.W., Chiu, C.-Y., Liu, Z.: Polycultural psychology. Annu. Rev. Psychol. **66**, 631–659 (2015)
30. Schwartz, S.H.: Rethinking the concept and measurement of societal culture in light of empirical findings. J. Cross Cult. Psychol. **45**(1), 5–13 (2014)
31. Wilson, A.: 'Mixed race' children in British society: some theoretical considerations. Br. J. Sociol. **35**, 42–61 (1984)
32. Phinney, J.S., Alipuria, L.L.: At the interface of cultures: multiethnic/multiracial high school and college students. J. Soc. Psychol. **136**(2), 139–158 (1996)
33. Rockstuhl, T., Ng, K.-Y.: The effects of cultural intelligence on interpersonal trust in multicultural teams. In: Handbook of Cultural Intelligence, pp. 206–220 (2008)
34. Pick, A.D., Pick Jr., H.L.: Culture and perception. In: Carterette, E.C., Friedman, M.P. (eds.) Handbook of Perception, pp. 19–39. Academic Press (1978)
35. Barsness, Z.I., Bhappu, A.D.: At the crossroads of culture and technology social influence and information-sharing processes during negotiation. In: Gelfand, M.J., Brett, J.M. (eds.) The Handbook of Negotiation and Culture, pp. 350–373 (2004)
36. Nisbett, R.E., Miyamoto, Y.: The influence of culture: holistic versus analytic perception. Trends Cogn. Sci. **9**(10), 467–473 (2005)
37. DePaulo, B.M., Friedman, H.S.: Nonverbal communication. In: Gilbert, D., Fiske, S.T., Lindzey, G. (eds.) Handbook of Social Psychology, 4th edn., vol. 2, pp. 3–40. Random House, New York (1998)
38. Kersten, G.E., Koszegi, S., Vetschera, R.: The effects of culture in anonymous negotiations: experiment in four countries. In: System Sciences, HICSS. IEEE (2002)
39. Zartman, I.W.: Prenegotiation: phases and functions. Int. J. **44**(2), 237–253 (1989)
40. Stein, J.G.: Getting to the table: the triggers, stages, functions, and consequences of prenegotiation. Int. J. **44**(2), 475–504 (1989)
41. Raiffa, H., Richardson, J., Metcalfe, D.: Negotiation analysis: the science and art of collaborative decision making. The Balknap Press of Harvard University Press, Cambridge (2002)
42. Roszkowska, E., Wachowicz, T.: Inaccuracy in defining preferences by the electronic negotiation system users. In: Kamiński, B., Kersten, G.E., Szapiro, T. (eds.) GDN 2015. LNBIP, vol. 218, pp. 131–143. Springer, Cham (2015). https://doi.org/10.1007/978-3-319-19515-5_11
43. Roszkowska, E., Wachowicz, T., Kersten, G.: Can the holistic preference elicitation be used to determine an accurate negotiation offer scoring system? A comparison of direct rating and UTASTAR techniques. In: Schoop, M., Kilgour, D.M. (eds.) GDN 2017. LNBIP, vol. 293, pp. 202–214. Springer, Cham (2017). https://doi.org/10.1007/978-3-319-63546-0_15
44. Roszkowska, E., Wachowicz, T.: The application of item response theory for analyzing the negotiators' accuracy in defining their preferences. In: Bajwa, D., Koeszegi, S.T., Vetschera, R. (eds.) GDN 2016. LNBIP, vol. 274, pp. 3–15. Springer, Cham (2017). https://doi.org/10.1007/978-3-319-52624-9_1
45. Swami, V., Tovée, M.J.: Female physical attractiveness in Britain and Malaysia: a cross-cultural study. Body Image **2**(2), 115–128 (2005)
46. Van Vianen, A.E., et al.: Fitting in: surface-and deep-level cultural differences and expatriates' adjustment. Acad. Manage. J. **47**(5), 697–709 (2004)

FITradeoff Method for the Location of Healthcare Facilities Based on Multiple Stakeholders' Preferences

Marta Dell'Ovo[1], Eduarda Asfora Frej[2([⊠])], Alessandra Oppio[3],
Stefano Capolongo[1], Danielle Costa Morais[2],
and Adiel Teixeira de Almeida[2]

[1] Department of Architecture, Built Environment and Construction
Engineering (ABC), Politecnico di Milano, via Bonardi 9, 20133 Milan, Italy
marta.dellovo@polimi.it
[2] CDSID - Center for Decision Systems and Information Development,
Universidade Federal de Pernambuco – UFPE, Av. Acadêmico Hélio Ramos,
s/n – Cidade Universitária, Recife, PE 50740-530, Brazil
eafrej@cdsid.org.br
[3] Department of Architecture and Urban Studies (DAStU),
Politecnico di Milano, via Bonardi 3, 20133 Milan, Italy

Abstract. Multiple stakeholders' preferences are considered for solving a healthcare facility location problem in the city of Milan, Italy. The preference modeling is based on the Flexible and Interactive Tradeoff (FITradeoff), a Multicriteria Decision Making (MCDM) method used to elicit criteria scaling constants in additive models. FITradeoff is an easy tool for decision makers, because it requires them to exert less effort than other traditional elicitation methods, as the tradeoff procedure. Therefore, it is expected that fewer inconsistencies will appear during the elicitation process. Sixteen criteria were used to evaluate in which of six potential areas a new hospital could be sited. An analyst with a strong background in MCDM interviewed four actors, and elicited their preferences with the help of the FITradeoff Decision Support System (FITradeoff DSS).

Keywords: Healthcare facilities location · Multicriteria decision-making Additive model · FITradeoff

1 Introduction

Selecting the most suitable area for siting healthcare facilities is not an easy task, for several reasons. First, it is a decision which has to consider the long-term consequences. For example, if an appropriate decision is made, it may well increase the hospital's advantages over its competitors [3]; secondly, site selection should take into consideration the surroundings and their influences on the site [20]; and finally, it is a multidimensional decision-making problem, which involves addressing multiple conflicting objectives [18].

© Springer International Publishing AG, part of Springer Nature 2018
Y. Chen et al. (Eds.): GDN 2018, LNBIP 315, pp. 97–112, 2018.
https://doi.org/10.1007/978-3-319-92874-6_8

Moreover, when the problem involves multiple actors, each seeking to exert their own influence on the decision process, it may become even more complex. According to de Almeida and Wachowicz [9], decisions in which multiple decision makers (DMs) are involved are more challenging compared to individual decisions, because in addition to the conflicting objectives considered, multiple actors have different viewpoints, preferences and aspirations. Due to the wide variety of actors that are usually involved in such decisions, facility location problems are widely explored in the literature as a Multicriteria Group Decision Making (MCGDM) problem.

In the context of a non-compensatory rationality, Norese [22] applied the ELECTRE method for a participatory decision-making process with 45 DMs who wished to locate a waste disposal plant in the District of Turin, Italy. The ELECTRE method was also applied by Hatami-Marbini et al. [16] in order to assess hazardous waste recycling facilities in an environmental context. Likewise, the PROMETHEE method has also been used for aiding site location problems. Ishizaka and Nemery [17] proposed a multi-phase approach based on PROMETHEE GDSS to assist facility location decisions, and Tavakkoli-Moghaddam et al. [27] used the PROMETHEE method for facility location problems in a fuzzy environment, with a new approach called Z-PROMETHEE. Fuzzy-based approaches are widely explored in the literature on locating facilities [1, 14, 18, 23].

As for the context of a compensatory rationality, Chou et al. [4] presented a fuzzy multi-attribute group decision-making model based on a simple additive weighting system for aiding facility location selection problems, which used both objective and subjective criteria. Preference modeling is one of the most critical issues in MCDM, especially within the scope of a compensatory rationality [7], because the information that the DM requires for the decision-making process may be time-consuming, tedious and difficult to provide [25]. This may therefore lead to inconsistent results. It is issues like these that prompt the use of partial/incomplete information methods. Approaches based on incomplete information have also been widely used for aiding group decision and negotiation processes [5, 13, 26].

Given this context, this paper undertakes a case study of a healthcare facility location problem in the city of Milan (Italy), in which multiple stakeholders are involved. The aim of this work is to show how preference modeling was conducted with the different actors by considering partial information in the context of a compensatory rationality through a Flexible and Interactive multicriteria method (FITradeoff method [8]). The perceptions of the different actors regarding the applicability of the method so as to find a compromise solution for the group as whole are also described.

This paper is organized as follows. Section 2 describes the context of the healthcare facility location problem. It also identifies the different actors involved, potential alternatives and the set of criteria that is used to evaluate this problem. Section 3 gives a brief overview of the FITradeoff method, and it also describes the preference modeling process which was carried out with multiple stakeholders. Section 4 analyzes and discusses the results obtained, and finally Sect. 5 presents the final remarks.

2   Healthcare Facility Location Problem

Locating healthcare facilities requires the participation of different levels and hierarchies of actors since the final decision will impact society as a whole, the natural environment and the built environment. In Italy, in fact, each region is responsible for promoting health policies, managing resources and dividing them equitably. Therefore, planning and implementing these policies has to attend to the wants and needs of a wide range of actors. First, there are political actors who should represent the will of citizens and respect their needs, to do which they may possibly need to stop proposals or, on the contrary support proposals. Secondly, while healthcare companies can be considered as being in the private sector and able to represent their own interests, yet at the same time, they promote public health. Other actors include: health facility users, health workers, regular patients, helpers etc.

Therefore, this paper focuses on the complex issues that arise from dealing with the analysis of the various stakeholders involved and how best to manage and solve the decision-making problem when different and conflicting actors are involved.

2.1   Analysis of Stakeholders

Since "the outcome of a decisional process depends on the actors" [12], it is only after the decision problem has been framed, that the people involved in the process can be identified. This will include specifying their role and in particular whose interests are involved. An important issue clarified by Dente [12] is that it is possible to consider a group of actors as an individual subject, with the power of influencing the decision problem. When the interaction among them is stable and everyone is free to express his/her own opinion without pressure, in fact, they act for the common good. Otherwise, if the situation analyzed considers a group of actors as individuals who have different aims and act for various reasons with a view to satisfying the interests of their own sectoral interest, it is more challenging to aggregate their opinions in order to achieve a common judgement. The suggestion in this case is to observe the situation in order to proceed empirically and understand the actors' behavior, but what is more important is to identify and describe their goals, and if these conflict with the goals of others, the group should make an effort to satisfy all goals.

The literature divides actors into categories in accordance with the resources they control [12]. *Political resources* consider people who are the parties of public policies and the resources are related to the consensus an actor is able to gain and the amount of these; the greater the consensus, the greater his/her influence and power on the final decision. Clearly, *economic resources* are related to the amount of money an actor is able to mobilize and this also depends on the client; in fact, the richer the client, the larger the amount of money that can be mobilized in order to acquire power. *Legal resources* belong to actors who have the power and the ability to act because of their administrative or legislative authority; the greater the legal behavior, the greater the possibility of achieving the right solution. *Cognitive resources* depend on the knowledge of the actor on that specific problem; the greater the amount of information held about the problem, the greater the actor's influence in solving it.

Since there are different resources and actors involved in the hospital site selection decision problem, the concept of complexity can be usefully re-introduced at this point as it is now easier to grasp what this means. In this problem, what can now be recognized is the complete the set of previously defined resources and who the stakeholders with specific interests in these are: political actors, bureaucratic actors, special interests, general interests and experts [11]. Each actor behaves and acts in accordance with the type of resources previously defined; political actors move political resources, bureaucratic actors move legal resources; actors with special interests move economic resources; actors with a general interest move both legal resources and cognitive resources; and experts move cognitive resources. Having classified the actors into five distinct categories, the analysis of the decision problem was carried out so as to identify for each category the corresponding actor in the real world is. In detail:

- *Political actors:* Health and Urban Assessors;
- *Bureaucratic actors:* Health and Urban General Manager;
- *Actors with special interests:* Local Health Unit Director;
- *Actors with general interests:* Common people; Non-Profit Organization (NPO), Non-Governmental Organization (NGO).
- *Experts:* Architects, Planners, etc.

A further analysis concerns the investigation of the level of interest and power of the five categories previously defined. In this phase, the actors are further classified into four classes according to their position in the matrix of power/interest, as can be seen in Table 1. In this paper, our analysis is carried out based on the classification of stakeholders presented below.

Table 1. Matrix of power vs level of interest. [23]

		Interest	
		Low	High
Power	Low	**Minimal effort** Nonprofit organizations; non-governmental organizations	**Keep informed** Common people
	High	**Keep satisfied** Bureaucratic actors, experts	**Key players** Political actors

Table 1 shows the hierarchy of power in this problem. The key players are the political actors. The power of the bureaucratic actors and experts is also high while that of the NPOs/NGOs and common people is low. As to the level of interest of the various actors in the project, this is high for the political actors and the common people.

2.2   La Città della Salute: Structuring the Problem

The case study chosen to test the methodological framework proposed in this article concerns the location of "La Città della Salute", Milan, Italy. The aim of the project is to create a single center that will promote research and cancer treatments, by relocating two existing hospitals: the Istituto Neurologico Carlo Besta and the Istituto dei Tumori. The idea to create an integrated public system of complex medical functions both of clinical and scientific excellence, a project which started around the year 2000, but the process for selecting the site has not been guided by specific criteria related to the demand for health services in a given catchment area, but only by economic and political reasons, which have been undertaken with very limited transparency. In fact, the site selected has been given for free by its private owner, who is from the Lombardy region – and in charge of public health policies and programs – provided that the costs of land reclamation are subsidized by the regional government. The reclamation works, which started in 2013, are still ongoing as only 25% of the site is now suitable for development (La Stampa Milano http://www.lastampa.it/2017/03/09/edizioni/milano/consegnato-il-primo-lotto-per-la-nuova-citt-della-salute-a-sesto-san-giovanni-zqBCVqijq6QZd5LuHsOOeM/pagina.html).

The process to frame a suitable set of criteria to solve the decision problem underwent a deep analysis of the state of art of existing evaluation tools that focus on assessing the energy performance of hospitals. This was supported by a review of the literature on research that has focused on this field.

In detail, this investigation sought to find common criteria and indicators that should be considered for projects on locating a site for healthcare facilities. A comparative analysis was performed to highlight and define the criteria that were the most used by some evaluation tools (LEED Healthcare, BREEAM Healthcare, Metaprogetto DM 12/12/00) [10] and cited in the literature analyzed. Moreover, from the analysis of the literature review, a convergence of criteria emerges, even if analyzed from different disciplinary perspectives and with different methodologies. A set of sixteen criteria was then defined by taking into consideration four different dimensions of the problem – Functional; Locational; Environmental and Economic – each of which was then related to appropriate criteria (Table 2).

Table 2. Set of criteria

Dimension	Criteria	Scale	Performance
Functional	C1. Building density	Continuous	Number of people living in and near the area
	C2. Health demand	0–100	Percentage of people above 65 years of age
	C3. Reuse of Built-up areas	0–1	Promotes the use of sites already exploited
	C4. Areas with the potential to become attractive	Continuous	Encourages the development of peripheral sites considering the distance from the city center
Locational	C5. Accessibility	Continuous	Volume of all public, private and sustainable transport or

(continued)

Table 2. (*continued*)

Dimension	Criteria	Scale	Performance
			infrastructure that reaches the site, and the number of parking places
	C6. Existing hospital	0–1	The presence could be an advantage or disadvantage according to the treatments provided by the new and the existing hospital
	C7. Services	Continuous	Number of specific facilities present in a radius of 800 m from the site
	C8. Sewer system	0–1	Presence of this infrastructure
Environmental	C9. Connection to green areas	0–4	Meets four characteristics, such as the possibility to reach parks and garden in a short time
	C10. Presence of rivers and canals	0–1	To avoid the choice of sites with hydraulic and hydrological instability
	C11. Air and noise pollution	0–4	Concentration of specific pollutants (PM, O_3, NO_2) and the dB level detected by surveys on site. The score is assigned according to the number of pollutants that exceeds the limits permitted by the national regulations and if the acoustic limit has been respected
	C12. Land contamination	High-medium-low	In line with the site being suitable to host a hospital
Economic	C13. Land size and shape	Continuous	Ratio between the dimensions of the site and those of the new hospital
	C14. Land ownership	High-medium-low	Related to the percentage of public and private areas. The presence of public owners is preferred since it shortens the overall time of the trade process. Low = public $\leq 33\%$; Medium = $33\% <$ public $< 66\%$; High = public $\geq 66\%$
	C15. Land cost	Continuous	Price €/sqm
	C16. Land use	High-medium-low	In relation to the tendency for the site to host new healthcare facilities in accordance with the type of land cover

Before Sesto San Giovanni was selected as the most adequate area for the location of "La Città della Salute", five other areas were proposed, but without being subjected to an analytical process or to a feasibility study. Thus, this paper will include the additional alternatives – Area 1; Area 2; Area 3; Area 4; and Area 5 – that correspond to the five additional areas in the city of Milan proposed by the Municipality to locate the new healthcare facility. Sesto San Giovanni was also considered in our analysis, represented by "Area 6". Figure 1 illustrates the location of the six potential areas in the map of the city of Milan.

Fig. 1. Potential location areas in the city of Milan

Table 3 shows the decision matrix with the performance of the alternatives in each criterion.

Table 3. Decision matrix

	C1	C2	C3	C4	C5	C6	C7	C8	C9	C10	C11	C12	C13	C14	C15	C16
Area 1	158025	45152	1	4500	21921	1	14	0	3	0	3	0,5	0,13	1	203,56	1
Area 2	145160	39938	1	3100	12016	0	19	0	2	0	1	0,5	0,34	1	244,7	1
Area 3	145160	39938	0	5200	5617	0	10	1	2	0	3	1	0,14	0	250,48	0,5
Area 4	168456	48693	0	7700	3711	1	11	0	2	1	2	0	0,33	1	59,93	1
Area 5	168456	48693	1	8500	14	1	13	0	1	0,5	3	0	0,23	1	163	0,5
Area 6	82154	20147	1	9800	6519	0	17	1	1	0	1	1	0,4	0	15,25	1

3 Preference Modeling with Flexible Interactive Tradeoff

Within the scope of a compensatory rationality, this section sets out to show how multiple stakeholders' preferences can be modeled by using the Flexible and Interactive Tradeoff method [20] for the healthcare facility location problem in the city of Milan. First, Sect. 3.1 gives a brief background to the FITradeoff method, and then Sect. 3.2 describes how the preference modeling was conducted with the representatives of the four categories of stakeholders, as described in Sect. 2.

3.1 FITradeoff Method

Let us consider a multicriteria decision-making problem with set of criteria $C = \{c_1, c_2, \ldots c_n\}$ and a set of alternatives $\{a_1, a_2, \ldots a_m\}$. In the scope of Multi-attribute Value

Theory - MAVT [19], the alternatives are scored straightforwardly by using an additive aggregation function:

$$v(a_i) = \sum_{j=1}^{n} k_j v_j(a_i) \qquad (1)$$

In (1), k_j is the scaling constant (or weight) of criterion c_j, and the value function of criterion c_j with respect to alternative a_i is $v_j(a_i)$, which is normalized in such a way that the best outcome is set to one, and the worst outcome is set to zero. However, eliciting values of the scaling constants k_j is one of the most challenging tasks in MCDM. The classical tradeoff procedure developed by Keeney and Raiffa [19] for eliciting weights has a strong axiomatic foundation [28], but it is not very often used because of the difficulty that DMs have in giving the cognitively demanding information requested in the elicitation process. According to behavioral studies, the traditional tradeoff procedure presents around 67% of inconsistency when applied [2], which can be explained by the high cognitive effort that DMs have to make when answering tradeoff questions.

Let us assume that criteria weights are ranked according to a DM's preference structure, in such a way that $k_1 > k_2 > \ldots > k_n$. In the classical tradeoff elicitation process, DMs are asked to compare adjacent criteria with a view to finding indifference relations. For example, let us assume there are two hypothetical alternatives, $H_A = (w_1, w_2, \ldots, x_j, w_{j+1}, \ldots, w_n)$ and $H_B = (w_1, w_2, \ldots, w_j, b_{j+1}, \ldots w_n)$., where w_j and b_j are respectively the worst and the best outcome of criterion c_j, and x_j is an intermediate outcome for criterion c_j. The DM is required to specify the exact value of x_j^I which makes H_A and H_B indifferent according to his/her preferences - for further details, see [25]. Equations obtained from such indifference relations form an equation system which can be solved in order to find the values of the scaling constants k_j, and thus the global value of eachlternative is calculated based on (1). Specifying such indifference points is a highly demanding cognitive task, which leads to a high rate of inconsistencies [8]. Prompted by these issues, the FITradeoff method was developed in order to improve the applicability of the classical tradeoff for DMs, with a less cognitively demanding process which may lead to a reduction in the inconsistency rate. In FITradeoff, the DMs do not need to specify exact indifferent points, but, instead, to state strict preference between hypothetical alternatives, which is a cognitively easier task. If the DM states that H_A is preferred to H_B for $x_j = x_j'$, then the following inequality is obtained:

$$k_j v_j \left(x_j' \right) > k_{j+1} \qquad (2)$$

On the other hand, for another value of $x_j = x_j'' < x_j'$, the DM may state that H_B is preferred to H_A, in such a way that another inequality is obtained:

$$k_j v_j \left(x_j'' \right) < k_{j+1} \qquad (3)$$

With these inequalities, a weight space φ is obtained. The inequalities act as constraints for linear programming problems (LPP) that are run at each step in an attempt to find a solution for the problem [8]. The elicitation process is interactive, and at each interaction the DM answers another question by comparing two hypothetical alternatives, which results in a new inequality of type (2) or (3) being obtained. The new inequality is incorporated as a constraint of the LPP, which is then run again. This process goes on until a solution for the problem is found. The DM may also interrupt the elicitation process before the end, if he is not willing to give additional information. Figure 2 summarizes the elicitation process conducted in FITradeoff.

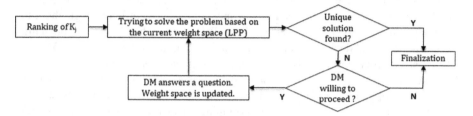

Fig. 2. FITradeoff elicitation process

The FITradeoff method is operated by means of a Decision Support System (DSS), which is available by request to the authors on the website www.fitradeoff.org.

The FITradeoff method can be applied for solving multicriteria decision making problems also when multiple decision makers are involved [6]. The elicitation process with the different actors can be conducted either simultaneously or independently. In the former, the DMs have to set their agenda so that they are simultaneously available, and thus the elicitation is conducted jointly with all of them. This approach has the advantage that the DMs may express their different points of view during the meeting, in such way that discussions are stimulated. In the independent elicitation, however, the interview with each DM is conducted separately, according to their own availability, within a deadline. If there is no common solution in the final subset of alternatives for all DMs, a final joint meeting may be necessary in order to make a final decision for the group. In the present case, the elicitation was conducted separately with each stake-holder, due to their limited availability.

3.2   Eliciting Stakeholders' Preferences

For each category of stakeholders – key players; keep satisfied; keep informed; and minimal effort -, one specific person responsible for representing his own class was chosen to be interviewed, and thus a total of four people had their preferences elicited. The idea was that all the classes of stakeholders had at least one representative in the elicitation process.

Interviewing the Actors. The category 'key player' was represented by an engineer working for the Lombardy region, who is Director of the investment system in the Department of the General Direction Welfare (DG Welfare). The competences of this sector concern the management of the Regional Healthcare Service; its tasks regarding the programming and monitoring of the health network; the coordination of the different levels of healthcare facilities, in the region; the promotion of wellbeing; and the prevention of disease.

For the category 'keep satisfied', an architect with competences in the field investigated by the research in [11] was selected. His background specifically includes studies on the design of hospitals, urban planning and the subject of real estate. The actor was defined as having high power since his opinion, guided by his experience and specific knowledge, can radically influence the decision problem.

As for 'keep informed', and in particular common people, a normal user of the hospital was chosen. He is a patient that normally goes to the hospital twice a year and to a doctor around six times per year. He does not have any competences or knowledge about the design of facilities nor on how to manage them. This category has low power but a high interest since the final decision could strongly affect their general well-being [15].

The last category analyzed is the minimal effort one, represented by NPOs and NGOs. An architect working in the field of the sustainable development was interviewed. He belongs to a cooperative of architects and engineers that used to work in close collaboration with associations in developing countries. In fact, his participation is relevant since he knows different contexts and how to work in several environments.

The four interviews were conducted individually, with no interference from other people. An analyst with a strong background in MCDM aided the decision process.

Interviews were divided into five main phases:

1. During the first phase, the analyst explained the decision problem, the case study that was to be undertaken, and the meaning of the sixteen criteria to the actor. The analyst took care that his explanations were objective and did not express his opinions.
2. The second phase was devoted to explaining the FITradeoff method. This included when it can be applied, for what purpose, what it consists of and what its potential is.
3. Having explained the basic concepts of the method and of the decision problem, the third phase focused on making clear to the actor what his role would be.
4. The fourth phase dealt with how to apply the FITradeoff method. This starts with the ranking of criteria weights, and then moves on to the question-answering step, until a unique solution is achieved.
5. During the fifth and last phase, the role of the analyst was to understand what the opinion of each actor was about the methodology applied, and the decision problem; to check overall awareness of the situation; and to get global suggestions for improvements to the process.

Results. The first step of the elicitation process in FITradeoff is the ranking of the criteria scaling constants, as shown in Fig. 2. The FITradeoff DSS allows the DM to choose whether he wishes to rank the criteria weights based on a holistic evaluation or by pairwise comparison. Since there are a large number of criteria, it is easier to conduct it by pairwise comparison in this case. The result of this step for each stakeholder is shown in Table 4.

Table 4. Ranking of criteria weights

Rank	Key player	Keep satisfied	Keep informed	Minimal effort
1	$k(C5)$	$k(C5)$	$k(C10)$	$k(C1)$
2	$k(C7)$	$k(C7)$	$k(C5)$	$k(C5)$
3	$k(C2)$	$k(C10)$	$k(C11)$	$k(C3)$
4	$k(C13)$	$k(C11)$	$k(C13)$	$k(C6)$
5	$k(C10)$	$k(C15)$	$k(C12)$	$k(C4)$
6	$k(C14)$	$k(C8)$	$k(C14)$	$k(C14)$
7	$k(C12)$	$k(C12)$	$k(C16)$	$k(C12)$
8	$k(C3)$	$k(C6)$	$k(C15)$	$k(C9)$
9	$k(C6)$	$k(C3)$	$k(C6)$	$k(C7)$
10	$k(C4)$	$k(C1)$	$k(C3)$	$k(C13)$
11	$k(C1)$	$k(C2)$	$k(C7)$	$k(C2)$
12	$k(C11)$	$k(C14)$	$k(C9)$	$k(C8)$
13	$k(C9)$	$k(C16)$	$k(C1)$	$k(C11)$
14	$k(C15)$	$k(C9)$	$k(C8)$	$k(C15)$
15	$k(C8)$	$k(C4)$	$k(C4)$	$k(C10)$
16	$k(C16)$	$k(C13)$	$k(C2)$	$k(C16)$

The second step was the question-answering process, in which the DMs interactively answered questions put by the DSS on comparing different criteria, so that the DMs could consider tradeoffs between them. After each question was answered, an LPP model is computed, with a view to finding potentially optimal alternatives (POAs) [8]. The interactive question-answering process continues until a unique alternative is found to be potentially optimal, which is the optimal alternative. The Director (*key player* actor) and the NPO architect (*minimal effort* actor) needed to answer only two questions until a unique solution was found. The architect with hospital design experience (*keep satisfied* actor) had to answer six questions in order to find a solution, while the common patient (*keep informed* actor) needed to answer thirteen questions. The final result was the same alternative for all the four actors: Area 1. As discussed in the next section, the flexibility of FITradeoff allows the *keep satisfied* and *keep informed* actors to find a solution at the second question, as the other two actors had done.

4 Discussion of Results

Whereas the *key player* and *minimal effort* actors came to a solution after answering only two questions, for *keep satisfied* and *keep informed* actors, a greater number of questions would be necessary by following the standard process. One of the key features of the FITradeoff DSS is its flexibility in the elicitation process. Moreover, it also allows difficult questions to be skipped in the process and tries to find a solution by means of a holistic evaluation of the remaining set of POAs. The analyst can do this at any time throughout the process. Let us consider this possibility when the *key players* and *minimal effort* actors have found a solution. That is, after two questions have been answered. Figures 3 and 4 show the partial results for the *keep satisfied* and *keep informed* actors at this point, respectively.

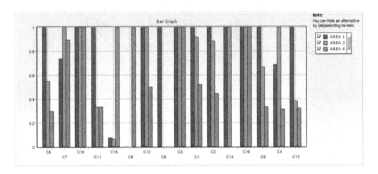

Fig. 3. Partial results for the *keep satisfied* actor after two questions

In Fig. 3, it can be noticed that Area 1 seems to have an advantage, as its performance is the highest in many criteria. On the other hand, it can be seen that it is too hard to choose one of the alternatives by analyzing Fig. 4.

Another flexibility of the DSS with this graphical visualization is indicated in the note at the top right-hand corner of the frame. That is, the user can select a small number of alternatives in order to make the holistic evaluation easier, as shown in Fig. 5. In this case, after analyzing the graph, an actor may correctly conclude that the performance of Area 1 is better than that of Area 3. It is important to note that the criteria with higher weights are on the left side.

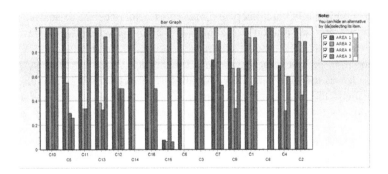

Fig. 4. Partial results for the *keep informed* actor after two questions

According to this first experience of testing the same decision problem with four actors, a correspondence in the final result can be highlighted. In fact, even with different rankings of criteria weights, the result as to which site was the most suitable one was Area 1. This can be also justified and understood since for all the actors one of the most important criteria is accessibility (C5), and the performance of Area 1 is much better than that of the other areas on this aspect, which creates a large discrepancy.

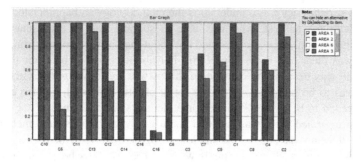

Fig. 5. Partial results for the *keep informed actor*, after selecting only two alternatives

Table 4 can be used to understand to which criteria the actors decided to assign most importance. The ranks defined by the interaction with the actors are varied and describe different wants, needs and personal preferences. It is possible, nevertheless, to see tendencies, similarities and differences. Criterion C5 (accessibility) always ranks among the first two positions. In fact, for the *key player* and the *keep satisfied* actors it is the most important one, while for the *keep informed* and *minimal effort* actors it is the second most important. Criterion C10 (presence of rivers and canals) is located among the first five positions for three of the four actors, but for the *minimal effort* actor it is one of the bottom-ranked ones, because he judged it as a prerequisite and irrelevant for the purpose of the research.

Some expectations, which were recorded before starting the interviews, were not satisfied from the results obtained. For example, it was expected that the actor representative of common people (*keep informed*) would give more importance to qualitative criteria, such as the presence of services (C7) or the connection to green areas (C9); on the contrary, he assigned more influence to technical criteria such as the presence of rivers and canals (C11) and air and noise pollution (C11). Instead, the actor who represented NPOs and NGOs (*minimal effort*) respected the expectation that assigned more importance to social criteria and this is also related to the issue of the sustainability, such as building density (C1) and the reuse of built-up areas (C3).

The interaction also allows us to have an overview of the perceptions of the interviewees and to ask them to make suggestions for improving the process. In fact, after the evaluation phase which was carried out with the support of the FITradeoff method, the last phase consisted of understanding the actors' opinions about the methodology applied and the decision problem.

The *key player* actor was in general satisfied with the application and considered that the process was easy to understand, but he was not satisfied with the ranking part because of the time it took to make the pairwise comparison. This can be easily explained since the number of criteria influences the number of questions made by the program to frame a ranking. The key player also suggested that the number of criteria could be reduced. He claimed not only that some of them were correlated but also that there were too many of them. He further affirmed that as they referred to different aspects of the problem, it would be better to group them in macro-areas.

The *keep satisfied* actor also understood the methodology and appreciated the direct interaction with the analyst, who was able to support him in each phase without expressing his opinion. He was concerned not about the pairwise comparison, which he considered was clear and easy to perform, but about choosing between two consequences. This was because he considered that it was too obvious for some of them which one was to be preferred. As to the set of criteria proposed, he also considered that these needed to be reviewed since some elements were redundant.

The *keep informed* actor was satisfied about the whole process and methodology. Although he does not have the skills and knowledge needed to make a formal analysis of the decision problem, he understands why the case study is being undertaken and why the choices he is asked to make are important, namely, they will contribute towards the final results. In other words, he is aware that his preferences could change the final decision and therefore he is also aware that his role is important.

The *minimal effort* actor found it easy to follow the different steps, but he criticized some of them. For example, with regard to the pairwise comparison, he stressed that in his opinion, it is not logically consistent to compare some criteria with each other because some criteria are non-comparable with others. Drawing on his past experiences of being involved in decision-making processes and in particular, in inclusive processes, he regarded the part of the software related to visualization was weak, because the visuals are designed to be interpreted by experts or technicians who have specific knowledge.

5   Conclusions

By observing the interaction with the actors and applying the method to a real world case study, the strengths and weakness of the methodology were identified. In fact, the extent to which the large number of criteria influences the number of questions during the pairwise comparison became apparent. Moreover, some criteria are strongly correlated to each other and therefore the analytical framework should be reviewed with a view to reducing the number of criteria. The second part of the FITradeoff method was considered to be effective and helpful, since the graphics support the decision actors in their efforts to better understand the questions; only one actor considered that the visuals were ambiguous. An important strength to emphasize concerns the role played by the actors in the process: they were aware of the importance of their roles and that their choices would influence the results.

In this case, all actors obtained the same optimal area, but there is still the issue of whether or not to aggregate the different outputs obtained, in case the results were different for each actor. Is it reasonable to aggregate them? Or is it better to keep them apart from each other and to consider the influence of the actor in taking the final decision? Since the decision problem that this paper investigates is a social problem [21], the final choice should be shared by the whole community in order to find mediation and therefore to try to satisfy different needs and expectations. In some situations, the elicitation process can be conducted simultaneously with all the actors, as previously mentioned in Sect. 3.1. In this case, the DMs can express their different points of view for the others, and perhaps they may be able to reach a common solution based on these discussions. In the present case study, however, the elicitation was conducted separately with each DM, and thus a possible solution in case the results were different for each actor could be a further interaction together with every representative of the categories defined, in order to find a common solution for the group as a whole [24]. Proceeding along this theoretical perspective, it may also be possible to involve a greater number of stakeholders in the final decision and to work towards more stakeholders participating throughout the process.

Acknowledgments. This study was partially sponsored by the Brazilian Research Council (CNPq) for which the authors are most grateful.

References

1. Bashiri, M., Hosseininezhad, S.J.: A fuzzy group decision support system for multifacility location problems. Int. J. Adv. Manuf. Technol. **42**, 533–543 (2009). https://doi.org/10.1007/s00170-008-1621-3
2. Borcherding, K., Eppel, T., Von Winterfeldt, D.: Comparison of weighting judgments in multiattribute utility measurement. Manag. Sci. **37**, 1603–1619 (1991). https://doi.org/10.1287/mnsc.37.12.1603
3. Chiu, J.E., Tsai, H.H.: Applying Analytic Hierarchy Process to select optimal expansion of hospital location: the case of a regional teaching hospital in Yunlin. In: 10th International Conference on Service Systems and Service Management (ICSSSM), pp. 603–606. IEEE (2013). https://doi.org/10.1109/icsssm.2013.6602588
4. Chou, S.Y., Chang, Y.H., Shen, C.Y.: A fuzzy simple additive weighting system under group decision-making for facility location selection with objective/subjective attributes. Eur. J. Oper. Res. **189**, 132–145 (2008). https://doi.org/10.1016/j.ejor.2007.05.006
5. Clímaco, J.N., Dias, L.C.: An approach to support negotiation processes with imprecise information multicriteria additive models. Group Decis. Negot. **15**, 171–184 (2006). https://doi.org/10.1007/s10726-006-9027-9
6. de Almeida, A.T.: FITradeoff method for resolving evaluation of criteria by interactive flexible elicitation in group and multicriteria decision aid. CDSID Working Paper also Presented as Keynote at Joint International Conference of the INFORMS GDN Section and the EURO Working Group on DSS, Toulouse (2014)
7. de Almeida, A.T., Cavalcante, C.A.V., Alencar, M.H., Ferreira, R.J.P., Almeida-Filho, A.T., Garcez, T.V.: Multicriteria and Multiobjective Models for Risk, Reliability and Maintenance Decision Analysis. International Series in Operations Research & Management Science, vol. 231. Springer, Cham (2015). https://doi.org/10.1007/978-3-319-17969-8
8. de Almeida, A.T., de Almeida, J.A., Costa, A.P.C.S., de Almeida-Filho, A.T.: A new method for elicitation of criteria weights in additive models: flexible and interactive tradeoff. Eur. J. Oper. Res. **250**, 179–191 (2016). https://doi.org/10.1016/j.ejor.2015.08.058
9. de Almeida, A.T., Wachowicz, T.: Preference analysis and decision support in negotiations and group decisions. Group Decis. Negot. **26**, 649–652 (2017). https://doi.org/10.1007/s10726-017-9538-6
10. Dell'Ovo, M., Capolongo, S.: Architectures for health: between historical contexts and suburban areas. Tool to support location strategies. Technè J. Technol. Arch. Environ. **12**, 269–276 (2016). https://doi.org/10.13128/techne-19362

11. Dell'Ovo, M., Frej, E.A., Oppio, A., Capolongo, S., Morais, D.C., de Almeida, A.T.: Multicriteria decision making for healthcare facilities location with visualization based on fitradeoff method. In: Linden, I., Liu, S., Colot, C. (eds.) ICDSST 2017. LNBIP, vol. 282, pp. 32–44. Springer, Cham (2017). https://doi.org/10.1007/978-3-319-57487-5_3

12. Dente, B.: Understanding policy decisions. In: Dente, B. (ed.) Understanding Policy Decisions, pp. 1–27. Springer, Cham (2014). https://doi.org/10.1007/978-3-319-02520-9_1

13. Dias, L., Clímaco, J.: ELECTRE TRI for groups with imprecise information on parameter values. Group Decis. Negot. 9, 355–377 (2000). https://doi.org/10.1023/A:100873961

14. Ertuğrul, İ.: Fuzzy group decision making for the selection of facility location. Group Decis. Negot. 20, 725–740 (2011). https://doi.org/10.1007/s10726-010-9219-1

15. Grad, F.P.: The preamble of the constitution of the World Health Organization. Bull. World Health Organ. 80, 981 (2002)

16. Hatami-Marbini, A., Tavana, M., Moradi, M., Kangi, F.: A fuzzy group ELECTRE method for safety and health assessment in hazardous waste recycling facilities. Saf. Sci. 51, 414–426 (2013). https://doi.org/10.1016/j.ssci.2012.08.015

17. Ishizaka, A., Nemery, P.: A multi-criteria group decision framework for partner grouping when sharing facilities. Group Decis. Negot. 22, 773 (2013). https://doi.org/10.1007/s10726-012-9292-8

18. Kahraman, C., Ruan, D., Doğan, I.: Fuzzy group decision-making for facility location selection. Inf. Sci. 157, 135–153 (2003). https://doi.org/10.1016/S0020-0255(03)00183-X

19. Keeney, R.L., Raiffa, H.: Decision Analysis with Multiple Conflicting Objectives. Wiley, New York (1976)

20. Kumar, S., Bansal, V.K.: A GIS-based methodology for safe site selection of a building in a hilly region. Front. Arch. Res. 5, 39–51 (2016). https://doi.org/10.1016/j.foar.2016.01.001

21. Munda, G.: Social Multi-criteria Evaluation for a Sustainable Economy. Springer, Heidelberg (2008). https://doi.org/10.1007/978-3-540-73703-2

22. Norese, M.F.: ELECTRE III as a support for participatory decision-making on the localisation of waste-treatment plants. Land Use Policy 23, 76–85 (2006). https://doi.org/10.1016/j.landusepol.2004.08.009

23. Rao, C., Goh, M., Zhao, Y., Zheng, J.: Location selection of city logistics centers under sustainability. Transp. Res. Part D Transp. Environ. 36, 29–44 (2015). https://doi.org/10.1016/j.trd.2015.02.008

24. Rockloff, S.F., Lockie, S.: Democratization of coastal zone decision making for indigenous Australians: insights from stakeholder analysis. Coast. Manag. 34, 251–266 (2006). https://doi.org/10.1080/08920750600686653

25. Salo, A.A., Hämäläinen, R.P.: Preference assessment by imprecise ratio statements. Oper. Res. 40, 1053–1061 (1992). https://doi.org/10.1287/opre.40.6.1053

26. Sarabando, P., Dias, L.C., Vetschera, R.: Mediation with incomplete information: approaches to suggest potential agreements. Group Decis. Negot. 22, 561–597 (2013). https://doi.org/10.1007/s10726-012-9283-9

27. Tavakkoli-Moghaddam, R., Sotoudeh-Anvari, A., Siadat, A.: A multi-criteria group decision-making approach for facility location selection using PROMETHEE under a fuzzy environment. In: Kamiński, B., Kersten, Gregory E., Szapiro, T. (eds.) GDN 2015. LNBIP, vol. 218, pp. 145–156. Springer, Cham (2015). https://doi.org/10.1007/978-3-319-19515-5_12

28. Weber, M., Borcherding, K.: Behavioral influences on weight judgments in multiattribute decision-making. Eur. J. Oper. Res. 67, 1–12 (1993). https://doi.org/10.1016/0377-2217(93)90318-H

Capturing the Participants' Voice: Using Causal Mapping Supported by Group Decision Software to Enhance Procedural Justice

Parmjit Kaur[1(✉)] and Ashley L. Carreras[2]

[1] Department of Strategic Management and Marketing,
Faculty of Business and Law, De Montfort University,
Hugh Aston Building, Leicester LE2 7BQ, UK
pkcor@dmu.ac.uk
[2] School of Business and Economics, Loughborough University,
Leicestershire LE11 3TU, UK
a.carreras@lboro.ac.uk

Abstract. This paper examines the way in which causal mapping, aided by group decision software, adheres to the tenets of procedural justice. Causal mapping workshops utilise a dual facilitation process that enables the participants' "voice" to be heard. We demonstrate how a causal mapping process of investigation surfaces authentic qualitative data by aligning the process of investigation with the principles of procedural justice as found in organisational justice literature. This is supported by a statistical analysis of the dimension of procedural justice using the responses of workshop participants.

Keywords: Causal mapping · Procedural justice · Focus groups

1 Introduction

This paper examines how certain processes and procedures that embody the dimensions of Procedural Justice are utilised during focus group activities, and the extent to which they develop more meaningful levels of engagement with participants. This cross disciplinary study investigates the use of a soft operations research technique, Causal Mapping [11], in workshops with focus groups, where the objective of the workshop is to elicit meaningful information related to decision making. This is studied in the context of both private sector organisations and the student body of a UK university. The paper will be of particular interest to readers involved with policy making at all levels, from a process of investigation perspective and the utility of the method employed.

In the focus group workshops described in this study, a dual (software and human) facilitation process is used that allows the facilitators to surface the underlying issues that groups feel are key to that particular workshop discussion. Causal Mapping allows participants [5] to raise the key issues of concern by inviting them for their thoughts on a key prompt question which is used to start the focus group session. This prompt question is pre-determined to reflect an important decision-making area for that group. In doing this we witnessed a significant amount of "open, honest" and "insightful" information emerging during the facilitated focus group process. This leads us to believe that the nature of these facilitated focus groups allowed a more authentic voice to emerge.

The primary aim of the paper is to emphasise the learning points from the process of investigation we have used, and how it helps to draw out more "authentic" meaningful, detailed qualitative data, from participants. This becomes possible, we argue, since the process of investigation used is procedurally fair and hence better able to capture the participants' voice.

The main body of this paper is in 6 parts. Initially an outline of the Causal Mapping methodology is presented, followed by an outline of the processes employed in the workshops. We then examine how this dual facilitation process is aligned with Procedural Justice Dimensions, including voice and treatment effects. Finally an exploratory statistical analysis of the links between dual facilitation process and procedural justice dimensions is presented, before concluding.

2   Causal Mapping in Focus Group Workshops

Causal Mapping has its roots in the Personal Construct Theory of Kelly [24] and has been developed most notably by Eden [11] amongst others. Causal Mapping was chosen for collecting and analysing qualitative data in the workshop with participants as it helps provide a coherent picture of a situation. Causal mapping is an approach from Problem Structuring Methods (PSMs) [30] which allows a "systematic understanding of the issue at hand" [31] as it can deal with the complexity of issues that are interrelated [29]. This intervention approach, when teamed with a software aided process, improves efficacy as the intervention tool serves as a means of recording the data generated. The construction and analysis of the maps created in the workshops provide an insight into the underlying structure of an issue and, as "participants are facilitated through the complexity using a structured transparent process…….. this has positive effects on the data captured" [31, p. 832].

The focus group workshops were run using a mobile laboratory of networked laptops, using Group Explorer® software combined with the Decision Explorer® tool. The process allows the gathering, structuring and analysis of qualitative information that develops in the workshops. It allows the user to work with a model of interlinked ideas using maps created from the participants' own understanding of the main prompt question. These ideas (concepts) are gathered anonymously with the participants individually inputting their ideas via the laptops. This safety of non-recourse at this initial stage is vital in preventing the need for self-censorship and allowing a more authentic voice to emerge. The facilitators conducting the workshops ensured that any qualitative data generated was directly inputted by the participants themselves, thus embodying the understanding of these particular groups. As noted in the conclusion, this can be understood as showing the participants "respect" in a desire to improve interactional justice [2].

3   Focus Group Workshop Process

The aim of the workshops was to elicit information which would be rich in meaning and understanding from the participants' own perspective, such that any underlying issues of importance they surfaced through the course of the workshop would be prompted solely by "their view of the world as they understood it". This is in keeping with the phenomenological aspect of focus groups of "seeking everyday knowledge" [4, p. 356]. The stages of the focus group workshops followed a process of gathering, clustering, rating, causal linking and laddering. We now explain each stage.

Gathering: Wide gather of ideas (known as "surfacing" of concepts): Individual participants anonymously input their own thoughts on an initial prompt question.

The Group Decision software draws the concepts inputted into the individual networked laptops together into one space for examination. Once inputting is completed, the ideas or "concepts" are projected onto a large computer screen to allow the participants to read all the ideas generated by the group. This helps prevent the "group think" or "social loafing" [21] effect at the outset.

Clustering: The group members pick out concepts that appear to be of a similar theme leading to "clusters" of ideas that are related in some way. This building of clusters allows content to be reduced to a manageable level as it allows the group to work on each theme sequentially, as each cluster is copied and examined in turn on a new separate screen view. The process then becomes more visible and the whole group is involved in structuring the understanding.

Rating: Ratings are applied to these clusters whereby each participant votes anonymously using the software, as to which cluster represents the most important one to examine first and a ranking of clusters is achieved as a result. This ensures equitability in the process and prevents more vocal participants dominating the workshop process path.

Causal linking: This is used to develop the group's own meaning and understanding of the map created. As the participants input the directional arrows, the process allows discussion between the members of the group and encourages agreement on how the links should be constructed, though consensus across the groups is not essential, with the facilitator playing a key role in ensuring all participants opinions are able to be heard. The directional casual links are used to connect ideas, such that one concept leads to/causes another, for example the link between concept 70 and concept 68 (see top middle Fig. 1) indicates that this student group felt that "more formative assessment with feedback" would lead in some way to them being able to "achieve expected academic qualification". A negative sign on a link indicates that the preceding concept inhibits the following concept. (see Fig. 1 concept 67 "group work doesn't always accurately reflect the ability of students", negatively affects concept 68, "students being able to achieve their expected academic attainment".)

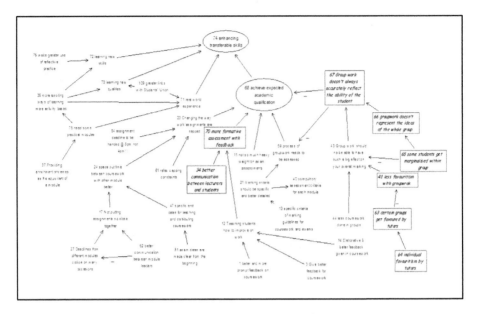

Fig. 1. Group One "Assessments" student map.

Laddering: By prompting the participants to consider why certain concepts matter to them we were able to work up towards goals (denoted by oval border statements) that they were hoping to achieve.

From the initial gathering of ideas (concepts) the groups worked together to build in "meaning" and understanding, such that the maps created directly reflected the negotiated understanding of the group. The role of the facilitator is to aid this process but at a "distance", helping only to structure and clarify the flow of the content not the actual content (concept statements) itself, whilst the maps are being "created" by the groups. To this end all the causal links (directional arrows) in the maps have been inputted by the participants. Thus it is fully their understanding that is represented in the structured maps. This approach is based upon Kelly's Personal Construct Theory [24] whereby a collection of ideas (concepts) and relationships (between the concepts) are connected in the form of a cognitive map. When these maps are utilised by groups they become referred to as causal maps. These maps help us to manage the content of a problem, whilst ensuring the social, political and process dynamics within the group are taken into account, so as to maintain a fair process [11–13]. During Causal Linking, it can be understood that Decision Explorer® is acting as a "dialectical tool, encouraging discussion and debate, helping people to explore the reasoning behind differing ideas held in the group" [1]. The maps represent a visual electronic memory of the discussion and can be added to and developed as the groups understanding evolves. The ability to add further content to the clusters, allows for the development of an emergent

Table 1. Participant feedback on dual facilitation process used.

Concept number from Student map	Statement in full
55	Visual representation of thinking on the screens helps to link understanding and give meaning
56	Process quicker than on paper and captures more content
57	We felt the questioning process was unbiased and not leading
58	Laptops were a familiar medium to use
59	Even with smaller a group the process yields a lot
60	Allowed us to discuss our experience as a whole
Concept number from Organization maps	Statement in full
82	Key benefit was keeping everything focused
83	Excellent presentation with clear and well explained conclusion
84	A follow on review breaking down further key points for each department
85	A very good way of collecting and assessing ideas in order to achieve a consensus of opinion, leading to action points
86	Good use of business model but reflecting on our actual business and it's requirements
87	Useful in clarifying goals/key issues for the company
88	Clear guidance for individual outcome and conclusion
89	Unlike typical "top-down" processes normally found in business. This allows equal participation by all parties
92	Excellent model for our team to understand our business

understanding in a seamless way over time. The notion of Concept Mapping [19] relies upon the view that abstract knowledge is more easily understood when transformed into visual representations. Thus it supports the use of a visual methodology to build understanding in the area of academic learning. Our feedback from participants in the workshop would appear to support this view (See Table 1 on feedback, concept 55). Maps are not provided here for brevity (and are available from authors)

A working model of the focus group sessions has been developed and can be represented by Fig. 2 below. The arrows in the looped feedback process represent the dual facilitation process in action. The right hand side of the diagram emphasises how the dual facilitation process encourages the display of desired extra role behaviours [25] in the workshops. The benefit of developing these behaviours for the quality of outcomes from the FGWs is discussed below.

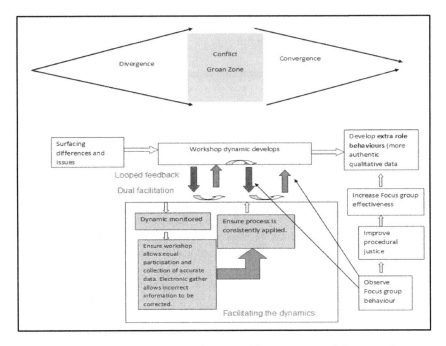

Fig. 2. Dual Facilitation Process of focus group workshops.

Figure 2 at the top provides the conventional notion of focus group activity passing through the three stages of divergence in group thinking, followed by the "Groan Zone" [22, 23] before the group moves into convergence of thinking. The modelling underneath developed by the authors represents the dual facilitation process as employed in the workshops. The initial surfacing of different ideas occurs in the gathering stage of the focus group process. As the workshop dynamic develops, the dual facilitation process allows a looped feedback process to operate whilst the workshop is operating live. Through the use of Group Explorer software, we monitor the group contribution to ensure an equal input of ideas from all of the participants. The dual facilitation process aids the enhancement of focus group outcomes as the workshop progresses, by encouraging the display of productive extra role behaviours.

In summary to this section the focus group examination undertaken tends to fit with the notion of experiential focus groups of the phenomenological type. In the context of the Experiential Focus Group as an effects application, [16] one looks to surface the "natural attitudes" of the focus group members. The primary focus of this phenomenological approach according to Calder [4] is to "draw out" the shared life experiences of the participants and is thus aimed at the "common-sense conceptions and ordinary explanations shared by a set of social actors".

We have not attempted to cover an overview of conventional focus groups, as this is well documented and referenced [6]. Other work in the field examines the notion of group think problems in focus groups [21], with others emphasising the importance of the group as a group notion, [7, 18].

The next section of the paper examines how this method of undertaking focus group activity, mirrors the dimensions of a "fair" process, as is discussed in procedural justice literature. The lack of research in this applied area is noted:

"The relatively small amount of research in group decision making is surprising considering its importance for both practice and theory. One possible explanation for this scarcity is the absence of an effective tool of for measuring fairness of procedures in a group context" [20, p. 386].

This section aims to draw the links between the facilitated, software driven focus group process used in the study and the characteristics of procedural justice.

4 Aligning the Dual Facilitation Process with Procedural Justice Dimensions

In organisational justice research concerns about fairness are based on the inter-related aspects of organisations, such as how resources are distributed - *distributive justice*; the fairness of decision making processes - *procedural justice*; the nature of interpersonal treatment received from others - *interactional justice* and collectively these justice dimensions are known as *organisational justice* [9]. Of these justice dimensions, the one which was the main aim of examination for the focus group work undertaken was *procedural justice,* since fairness of process is expected to enhance the focus group outcomes, in terms of levels of firstly; meaningful engagement with the process and secondly; the richness and authenticity of the qualitative data generated.

The work on Justice Literature has developed in waves with each dimension receiving prominence in certain decades; distributive (1950–1970), procedural (mid 1970s to mid-1990s), integrative (mid 1980s to present). Increasingly when examining the area of organisational justice, there has been a movement away from "distributive justice" to "procedural justice" (PJ) concerns.

The aspect of Justice in organisational literature is a subjective notion of justice that states that certain process and procedure types can enhance fairness judgments [28, p. 3]. "Procedures can refer to official rules of how things are done, how decisions are made etc. This represents the traditional view which in this study we refer to as Procedural Justice Narrow (PJN). An alternative and possibly more inclusive understanding of procedures can comprise all processes and interactions that occur in the context of organisational life" [3, p. 123], which we refer to as Procedural Justice Wide (PJW).

The quality of the sessions is indicated by the authenticity of the data generated and the number of concepts/statements that the participants volunteered in the session. We look to understand how our process ensures PJ as part of the process itself and not as a way of enabling PJW. That is: How does it ensure PJN?

There is a further distinction in the literature that will help our understanding. Organisational Justice can be seen to operate at two distinct but potentially interrelated levels. The individual self-interest models that state that participants are interested in fairness purely from improving their individual outcomes [26, p. 493], and the group oriented models which reflect the concerns of all the group members and are thus more complex in nature [28, 33, 34]. As we were using a group process that did not involve the participants making decisions that would necessarily directly affect their individual outcomes, it is argued that the group orientated models are more appropriate in framing this examination and this will be discussed further below under treatment issues in PJ.

In the area of PJ the work of Thibaut and Walker [32] paid particular attention to the "level of control" the participants believed they had in a process and the subsequent decisions arrived at through that process. They noted that participants reported higher levels of satisfaction when the process was seen as fair and as such even second best final decisions could be accepted by the participants so long as they had experienced control and fair participation in the earlier, process stage. [8, p. 426] "disputants viewed the procedure as fair if they perceived that they had process control" (that is, control over the presentation and sufficient time to present their case). This process control effect is often referred to as the "fair process effect" or "voice" effect [17, 28]. In this context fair decision making would allow participants control over the procedures that determine the outcome, as opposed to the outcomes themselves.

Linking this to our work, in an organisational context with a hierarchical structure such as a university or private sector organisation, direct decision making tends to reside at the top, and given that participants recognise this as the correct legal structure, they are hence prepared to accept "indirect opportunities" to impact on decision making as acceptable. This indirect aspect is termed "process control" by Thibaut and Walker [32], or the opportunity to express "voice". The process used allowed all participants to directly input their concepts (thoughts) into the Group Explorer system, without any censoring of views; hence we propose that the power to express "voice" for the participants is greatly enhanced by this process. This can be positively detected in the feedback from participants provided in Table 1, shown by concept 89.

Colquitt et al. [8] note that Leventhal broadened the determinants of procedural justice to points beyond process control [27]. This requires six criteria to be met if procedure is to be perceived as fair [8, p. 426]. These six determinants are drawn out to compare to the characteristics of the dual facilitation process used in the study in Table 2 below.

Table 2. Colquitt et al. [8]: p. 426.

Determinant; Colquitt et al.	Workshop process; Dual facilitation
(a) Procedures should be applied consistently across people and across time	We conduct the workshops using laptop laboratory setting, ensures uniformity over time, with the same prompt question for each group and general steps followed
(b) Procedures should be free from bias (i.e. ensuring that a third part has no vested interest in a particular settlement)	As facilitators we are independent of the university senior executive/organisation, and cannot impact on policy formulation at senior level
(c) Procedures should ensure that accurate information is collected and used in making decisions	Electronic gather of statements/concepts directly from the participants, ensures accurate collection of qualitative/experiential data with a clear audit trail through cluster building

(*continued*)

Table 2. (*continued*)

Determinant; Colquitt et al.	Workshop process; Dual facilitation
(d) Procedures should have a mechanism to correct flawed or incorrect decisions	The process can be used iteratively to ensure accuracy of information gathered. Concepts entered can be corrected electronically if incorrect
(e) Procedures should conform to prevailing standards of ethics or morality	Trained independent facilitators ensure process is ethically used with a correct employment of group norms in the focus sessions
(f) Procedures should ensure opinions of various groups affected by the decision have been taken into account	Students (and lower management in private organisations) are often not directly consulted in policy formulation, yet this process affords them a clear and transparent voice

When processes of investigation are embodying PJ determinants, the participants show commitment to the decisions made and will exhibit extra-role behaviours [25]. PJ also enhances the levels of voluntary contribution by "invoking the side of human behaviour that goes beyond the out-come driven self-interest" in exhibiting the extra-role behaviours [25]. All participants in our study were either volunteers or had been invited to take part by the lead member of the focus group (for the private organisations). Hence participants looked to experience a "fair" process of focus group investigation so as to engage meaningfully. In this study, the extra-role behaviour would be to divulge information that participants are not normally obliged to divulge and in doing so show "honesty" of opinion in a transparent manner. This would enable them to volunteer sensitive individual information (given the initial anonymity of the facilitated software driven process) relating to; how they felt they had been treated, how they had interacted with university tutors or to what extent as junior managers they could impact on decision making. As the inputting is anonymous electronic inputting to individual PCs, participants are less likely to self-sensor and will be more likely to engage in exhibiting extra- role behaviours and allow to surface sensitive individual opinions, which otherwise they would not feel safe to express. This can be illustrated by some concepts drawn from the maps, such as 64 in Fig. 1, "individual favouritism" which leads to 63 "certain groups get favoured by tutors", or statement 92; "students should be judged on ability rather than perception of ability", taken from another map not shown here. Similarly in the private organisation focus groups (given in Fig. 3) we had statements such as 29 "better dialogue between sales, warehouse and administration to achieve consensus agreement", clearly indicating communication issues that needed resolving.

The construction of the maps enables the facilitators to understand the conversation so that they may help surface more meaningful qualitative data. The understanding of qualitative research employed in this study is derived from Eisenhardt and Graebner [15] who stipulate that "qualitative research is highly descriptive, emphasising the

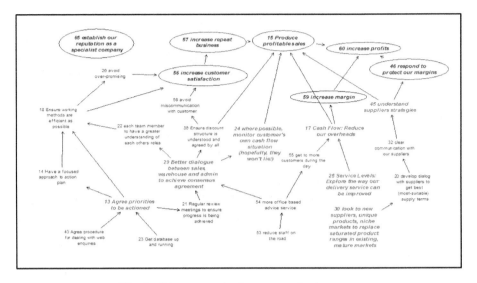

Fig. 3. Private organisation communication map.

social construction of reality, and focuses on revealing how extant theory operates in particular examples" [15, p. 28].

To understand how we are enabling PJN in our process, which in turn may enhance PJW we need to consider a more recent refinement of PJ terms.

4.1 Treatment Issues in Procedural Justice

The group engagement model of Tyler and Blader [34] gives a prominent role to procedural justice and is used to contextualise the work in this study. Within their model *treatment issues* are examined – participants value PJ (operationalised by voice or process control) because it aids the decision maker's ability to make equitable judgments. In this post 1990s examination of PJ more attention is given to the *interpersonal aspects* of procedures. This recognises that any process or procedure used in a group context will be a setting where participants are involved in social interaction, and is known as the *treatment aspect*. Interpersonal experience can range from being polite, rude, respectful and with hostility. The process used in these workshops is proposed to exhibit interpersonal fairness as one of the key functions of the independent facilitators is to ensure that the group conducts itself in a way that reinforces interpersonal fairness positively. The workshop sessions open with a slide on 7 "workshop conventions and norms" of operation that are adhered to throughout the workshop.

This shift in PJ, from a focus on decision making to interpersonal treatment aspects, shows the current development of PJ literature. It increasingly emphasises "pro-social outcomes, such as how to build trust, encourage responsibility and obligation, generate intrinsic motivation and stimulate voluntary co-operation with others" [33].

From the understanding of procedural justice dimensions, described in the section, a questionnaire has been constructed with 22 questions relating directly to these dimensions. Each question includes a 7 point Likert scale with a single dimension of agreement ranging from strongly disagree to strongly agree. Some questions were negatively worded to check for consistency of responses and thus detect any potential measurement error. These scales were reversed for the analysis. The questionnaire was given to 62 respondents across 7 workshops of various sizes ranging from 5–10 participants. The questionnaires were designed to check if the participants perceived the dual facilitation process was procedurally fair. The results from these questionnaires are discussed next.

5   Statistical Analysis of Procedural Justice Dimensions

The data was first checked for reliability of scales across the questions. We did this by examining the questions designed to capture fair procedures (12 questions) and treatment effects (4 questions) separately and using Cronbach's alpha. Of the remaining 6 questions 3 related to outcomes from the workshops and 3 to elements specific to Causal Mapping. For fair procedures we had a Cronbach's Alpha of 0.749 with the deletion of any of the items not significantly improving the result. For treatment effects the Alpha score was 0.835. Both of these are seen as indicating a high degree of correlation between the items, which suggests that it is appropriate to use scale reduction techniques. The Cronbach's Alpha also allows for the possibility of sub-dimensions within each of the dimensions of process and treatment effects (Cortina 1993) [10].

We proceeded to examine each of the dimensions in turn using Exploratory Factor Analysis, using a Principal Components Analysis (PCA) technique. For the treatment effect questions we had a Kaiser-Meyer-Olkin Measure of Sampling adequacy (KMO) of 0.704 and Bartlett's test of sphericity with a significance level below 1% indicating that there is a sufficiently high degree of correlation between the 4 items for PCA to be applicable. PCA yielded a single dimension with the single component accounting for 68% of the variation with 4 items.

In analysing the 12 items for process we used PCA and included or removed items on the basis of whether or not the KMO figure was suitably high, that all the of the items had an anti-image correlation above 0.5 and that the Bartlett's test of sphericity had a significance level below 0.05. We then inspected the components and used orthogonal rotation (varimax) to surface more clearly separated components, this approach assumes no correlation between the components or sub dimensions.

This process yielded a final rotated components matrix of with two components that accounted for 60% of the overall variation within the scales. The KMO was 0.668 with a significance level below 1% for Bartlett's test of sphericity.

The 7 remaining items loaded upon the two components with the following groups of questions:

Component 1 included: The participants were not able to make an equal contribution (reversed item contributionRev); The participants were treated equally in the process (equal); The process allowed for the group to discuss their concerns in an ethical manner (ethics); The opinions of the group were taken into account during the process (group opinions).

Component 2 included: I was able to contribute to the workshop without feeling the need to self-censor my contribution (censor); I felt able to express my ideas during the process (express); I think my opinions were being captured in the process (my opinions).

Component 1 appears to bear a strong resemblance to a notion of interactional fairness within the fair process effect outlined above and component 2 seems aligned to the notion of "voice" within the area of process control. It would appear that there is some prima facie evidence that the dual facilitation process using group software embodies the key dimensions of procedural justice incorporating the dimensions of voice, interactional fair processes and treatment. This supports the claims of Eden, Ackermann and Page [14] who have sought to incorporate the four-component model of Tyler and Blader [33] into their strategic decision making process using their "Journey" making approach.

A larger question is whether the existence of these dimensions delivers a better outcome for the participants. We have measured the participants' views on their satisfaction with the workshops and the perceived effectiveness of the workshops, these scores came out uniformly high. It is, therefore, difficult for any testing to pick up any correlation between the existence of the components above and variation in outcomes. In order to tease out any such relationships we would recommend that a larger data set is gathered, and a comparison with other approaches to running focus group workshops in undertaken. This would allow one to see if superior outcomes can be attributed to one approach when compared with another.

6 Conclusion

The examination here is firmly "practice based" in terms of context, as the study is drawn from focus group research on students and organisations.

The electronic gathering of qualitative statements (known as concepts) on the prompt question allows the participants a clear "voice". It has been noted that "voice" has value beyond is ability to shape decision making processes and outcomes [34, p. 351]. In the field of organisational research, justice is considered to be socially constructed. Although it is based upon a small sample and more comparative studies are needed, the statistical analysis appears to confirm the alignment between the dual facilitation process used in the Focus Group Workshops and the dimensions of procedural justice.

References

1. Banxia Reference Manual, Banxia software, Kendal (2002)
2. Bies, R.J., Moag, J.S.: Interactional justice; communication criteria of fairness. In: Lewicki, R., Sheppard, B., Bazermann, B.H. (eds.) Research on Negotiations in Organizations, vol. 1, pp. 43–55. JAI press, Greenwich (1986)
3. Blader, S.L., Tyler, T.R.: What constitutes fairness in work settings? a four- component model of procedural justice. Hum. Resour. Manag. Rev. **13**, 107–126 (2003)

4. Calder, B.J.: Focus groups and the nature of qualitative marketing research. J. Mark. Res. **14**, 353–364 (1977)
5. Carreras, A.L., Kaur, P.: Teaching problem structuring methods: improving understanding through meaningful learning. INFORMS Trans. Ed. **12**(1) 20–30 (2011). http://ite.pubs.informs.org/
6. Catterrall, M., Maclaren, P.: Focus groups in marketing research. In: Belk, R.W. (ed.) Handbook of Qualitative Methods in Marketing. Edward Elgar, Cheltenham (2006)
7. Chrzanowska, J.: Interviewing groups and individuals in qualitative market research. Sage, London (2002)
8. Colquitt, J.A., et al.: Justice at the millennium: a meta-analytic review of 12 years of organisational justice research. J. Appl. Psychol. **86**(3), 425–445 (2001)
9. Colquitt, J.A., et al.: What is organizational justice? a historical overview. In: Colquitt, J.A., Greenberg, J. (eds.) Handbook of Organizational Justice, pp. 3–56. Lawrence Erlbaum Associates, Inc. Hillsdale (2005)
10. Cortina, J.M.: What is coefficient alpha? an examination of the theory and applications. J. Appl. Psychol. **78**, 98–104 (1993)
11. Eden, C.: Cognitive mapping. Eur. J. Oper. Res. **36**, 1–13 (1988)
12. Eden, C., Ackermann, F.: Making Strategy: The Journey of Strategic Management. Sage, London (1998)
13. Eden, C., Ackermann, F.: Group decision and negotiation in strategy making. Group Decis. Negot. **10**, 119–140 (2001)
14. Eden, C., Ackermann, F., Page, K.: Strategic Management as Social Process in "Making Strategy". Chap. 2. Sage Publications, Thousand Oaks (2005)
15. Eisenhardt, K.M., Graebner, M.E.: Theory building from cases: opportunities and challenges. Acad. Manag. J. **50**(1), 25–32 (2007)
16. Fern, E.F.: Advanced Focus Group Research. Sage Publications, London (2001)
17. Folger, R., Cropanzano, R.: Organizational Justice and Human Resource Management. Sage publications, Thousand Oaks (1998)
18. Gordon, W.: Good Thinking: A Guide to Qualitative Research. Admap Publications, Henly-on-Thames (1999)
19. Hay, D., Kinchin, I., Lygo-Baker, S.: Making learning visible: the role of concept mapping in higher education. Stud. High. Educ. **33**(3), 259–311 (2008)
20. Jacobs, E., et al.: Of practicalities and perspective: what is fair in group decision making? J. Soc. Issues **65**(2), 383–407 (2009)
21. Janis, I.L.: Group Think: Psychological Studies of Policy Decisions and Fiascos, 2nd edn. Houghton Mifflin, Boston (1982)
22. Kaner, S.: Promoting mutual understanding for effective collaboration in cross-functional groups with multiple stakeholders. In: Schuman, S. (ed.) The IAF Handbook of Group Facilitation: Best Practices from the Leading Organisation in Facilitation. Jossey-Bass, San Francisco (2005)
23. Kaner, S.: Facilitator's Guide to Participatory Decision Making. Jossey-Bass, San Francisco (2007)
24. Kelly, G.A.: The Psychology of Personal Constructs: A theory of personality. Norton, New York (1955)
25. Kim, W.C., Mauborgne, R.A.: Procedural Justice, strategic decision making, and the knowledge economy. Strateg. Manag. J. **19**(4), 323 (1998)
26. Kovonosky, M.A.: Understanding procedural justice and its impact on business. J. Manag. **26**(3), 489–563 (2000)
27. Leventhal, G.S., et al.: Beyond fairness: a theory of allocation preferences. In: Minkula, G. (ed.) Justice and Social Interaction, pp. 167–218. Spinger, New York (1980)

28. Lind, E.A., Tyler, T.R.: The Social Psychology of Procedural Justice. Plenum, New York (1998)
29. Pidd, M.: Tools for Thinking: Modelling in Management Science. Wiley, Chichester (1996)
30. Rosenhead, J., Mingers, J.: Rational Analysis for a Problematic World Revisited. Wiley, Chichester (2001)
31. Shaw, D.: Journey making group workshops as a research tool. J. Oper. Res. Soc. **57**, 830–841 (2006)
32. Thibaut, J., Walker, L.: Procedural Justice. Lawrence Erlbaum Associates Inc., Hillsdale (1975)
33. Tyler, T.R., Blader, S.L.: Cooperation in Groups: Procedural Justice, Social Identity, and Behavioural Engagement. Taylor & Francis Group, Philadelphia (2000)
34. Tyler, T.R., Blader, S.L.: The group engagement model: procedural justice, social identity, and cooperative behavior. Pers. Soc. Psychol. Rev. **7**(4), 349–361 (2003)

The Effects of Photographic Images on Agent to Human Negotiations: The Case of the Sicilian Clan

Rustam Vahidov$^{(\boxtimes)}$

Department of Supply Chain and Business Technology Management,
Concordia University, Montreal, Canada
rustam.vahidov@concordia.ca

Abstract. Past studies in agent negotiations against humans have revealed important insights. The current study examines the effects of including a person's photograph as a proxy for an agent in negotiations on the outcomes. To this end an experiment employing photos of a young person, a mature person, and an older individual to represent agents was conducted. The results revealed significant difference in terms of the agreement utilities achieved by the three groups of agents. Therefore, the choice of an image for representing a software agent plays an important role in determining the outcomes of human vs. agent negotiations.

Keywords: Software agents · Electronic negotiations · Experimental studies

1 Introduction

Electronic negotiations allow parties to search for mutually beneficial deals over the Internet. Software agents could help businesses streamline negotiation processes by improving negotiation performance and consistency. Past studies have examined the viability and effectiveness of human vs. agent negotiations [1, 2]. It was found that, overall seller agents employing time-dependent tactics outperform humans performing the same role in negotiations with human buyers, and that task complexity mediates the performance of different agent tactics.

The current study investigates whether using a photographic image as a proxy for an agent has any impact on the outcome of negotiations. More specifically, it examines the influence of including an image of a younger, a mature, and an older male person on the negotiation performance. Photographic images of three actors featuring in an old classic French film "The Sicilian Clan" were included in the study. The film was released in 1969, and there is little chance that the subjects, who were recruited from the students of a large North American university, would recognise their "counterparts".

The results revealed that there was a significant difference in utilities achieved among the three groups of identically configured agents that used three different pictures to represent themselves. In particular, it was found that agents "wearing" a picture of a mature person achieved agreements with significantly higher utilities than other

© Springer International Publishing AG, part of Springer Nature 2018
Y. Chen et al. (Eds.): GDN 2018, LNBIP 315, pp. 127–135, 2018.
https://doi.org/10.1007/978-3-319-92874-6_10

agents did. What is more interesting, among the three groups of buyers there was no significant difference in the utilities of agreements.

Additionally, the subjects were given a questionnaire measuring their satisfaction with the counterpart and the achieved outcomes. There was no statistically significant difference between the subjects' assessments of the three agent categories, although simple comparisons of the results reveal interesting implications.

2 Related Work

Prior work studying the impact of photo images in agent vs. human negotiations has been scarce. There have been studies in incorporating images and photos in online commercial sites, using synthetic and anthropomorphic images in Human-Computer interactions, and providing social and visual cues in negotiations.

Being able to understand and predict the other party in e-commerce interactions tends to increase trust, as pointed out in [3]. This can be achieved by increasing social presence in commercial websites, in particular, by including photos of people. One study has examined the influence of the presence of photographs and videos in commercial websites on trust [4]. The findings suggested that presence of an image of a customer service representative enhances customers' trust in a website. This effect was stronger among Eastern culture subjects, as compared to the Westerners.

A study examining effects of anthropomorphic images on the perception of presence has been reported in [5]. The subjects were told they were either interacting with a human via an avatar, or with an agent. The task included getting to know their partner who may work with them in the future on a scavenger hunt for software technologies on the web. The findings indicated that having an anthropomorphic image tends to increase the perception of telepresence as compared to having no picture, however the appearance of the image plays a major role. Similar findings have been reported in a study investigating the effects of including 3D avatars in e-commerce websites on the feeling of telepresence [6].

In [7] it was postulated that facial features play an important point in the assessment of trustworthiness of a partner. They influence the way people perceive an unknown individual to be capable of reciprocal action. Using the trust game the authors have found that subjects were willing to invest more money when their partners had more trustworthy facial features.

In [8] a negotiation game was used to assess the influence of facial expressions on the subjects' perceptions. In particular, it was reported that "happy" facial expressions induced higher friendliness impressions. In the context of e-negotiations, one study compared negotiators credibility in online vs. face-to-face settings [9]. It was found that e-negotiators perceived their opponents to be less credible, as compared to the face-to-face group. It seems logical that including an opponent's picture in an online negotiation system could help raise the credibility in online negotiations.

In a Japanese study authors found that so called "Kawaii" appearance (implying cuteness) has an impact on the impression in negotiations [10]. Three images with "Kawaii", "normal" and "non-Kawaii" faces were shown to the subjects along with negotiation situations. It was also found that use of language may worsen the impression of "Kawaii" and improve the perception of "non-Kawaii" parties.

In [11] the impact of agents' expression of emotions on the opponent's behavior has been studied. The subjects were paired up with agents conveying anger, neutrality, or happiness during negotiations using verbal and non-verbal expression mode. The "angrier" agents achieved larger concessions from their opponents. In [12] authors reported that facial structure of men, specifically width-to-height ratio, can be used to predict negotiation behavior. In particular, men with higher ratio tend to be less cooperative.

The above work suggests that there could be an important influence of incorporating photo images in agent-to-human negotiations. The subsequent sections report our experimental setup and findings when using different male images in the process of offer exchange.

3   Experimental Setup

We have used a newly developed system for agent-to-human negotiations for the experiments. The experimental task included negotiations about a mobile plan with five issues: Price, Regular Air Time, Extra Air Time, Text Messaging, and Data. The agents were configured as being slightly competitive.

The competitiveness level for an agent has been specified using its utility curve. The curve dictates how an agent drops its utility over time. Making smaller concessions in the beginning is considered as a competitive or "greedy" tactic, as an agent is trying to get the better deals until the deadline starts approaching. The utility curve used by the agent in the experiment is depicted on Fig. 1.

Human users had to configure their preferences regarding the issues by themselves, as shown on Fig. 2. Then they were asked to indicate their preferences within each issue.

The agents' preferences were configured identically for all three groups of agents. One group was assigned a photo image of a younger person (Al), the other one an image of a mature person (Lino), and the third one of an older individual (Jean). Figure 3 shows the screen with the pictures assigned to the agents. The images included photographs of key actors who featured in a classical French film "The Sicilian Clan". The film was chosen because it featured three generations of actors, and also to minimize the risk of subjects recognizing the persons on the photos: the film is old and of European origin, while the subjects are North American undergraduate students. Also, the pictures shown together with offers are of small format, thus revealing little facial details other than age.

The negotiations were configured to run in a synchronous mode, whereby the subjects were asked to stay online and logged on until the end of the experiment. The maximum time assigned for the offer exchange was set at 30 min. Subjects were recruited among senior level students at a large North American University registered for "Business Application Development" classes. 159 invitations were sent to the subjects, of which 93 responded.

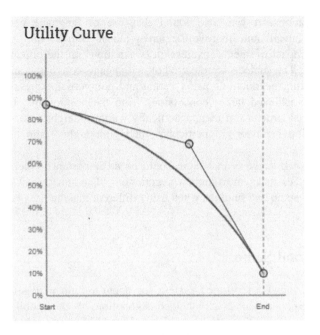

Fig. 1. Configuration of preferences

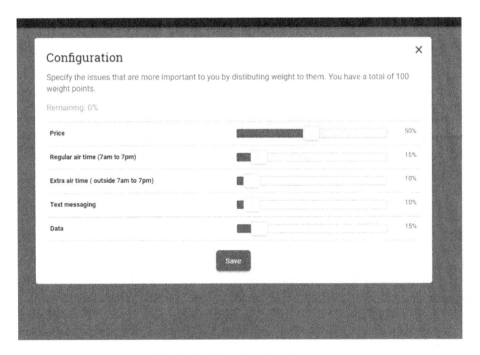

Fig. 2. Configuration of preferences

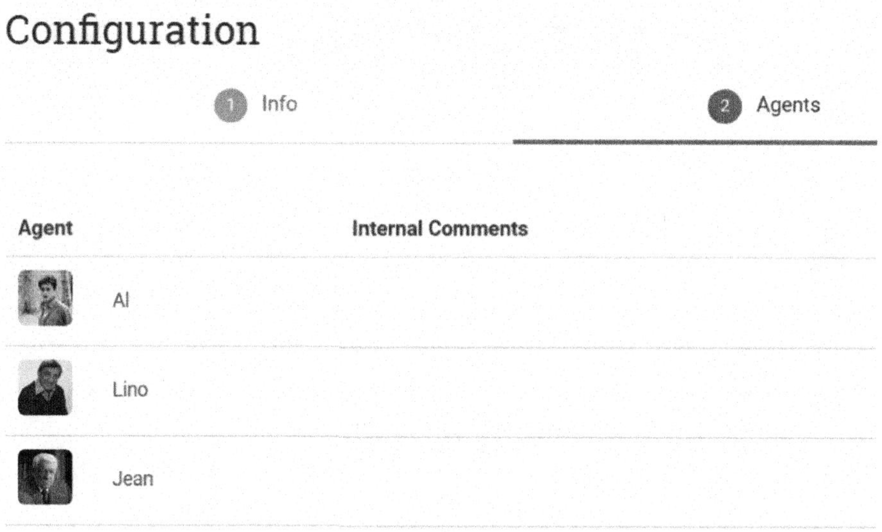

Configuration

Fig. 3. The images used in the experiment.

Subjects were given detailed instructions on how to use the system, and how to perform an experimental task. They were told that at any point they could terminate negotiations, if they thought that would be the best decision. They were randomly assigned to one of the three types of agents during the experimental period. During the offer exchange a counterpart's small picture was shown with each incoming offer. An example of subject-agent interaction is shown on Fig. 4.

4   Results

Of the 93 participants' negotiation records 75 were preserved, and the rest discarded, because the participating subjects accepted the first offer made by the agent counterpart, and, thus did not engage in true negotiations. Table 1 shows the agreement rates achieved by the three groups of agents. As one can see, the majority of negotiations ended with an agreement. The low number of no-agreement cases does not allow to compare agreement rates for statistical significance, though numerically the agents with mature person's photo have had highest rate.

In terms of the average number of offers, the "mature" agent had fewest (6.7), followed by the "older" (9.7) and "younger" (10.3) agents. The results of ANOVA showed that the difference was significant with a p-value of 0.029. Thus, the "mature" agent seems to be more "efficient" than the other two.

For the comparison of the utilities of agreement we have only included cases in which an agreement was reached. Additionally, few subjects had not correctly specified their preferences (e.g. having utility levels over 100), and these cases had to be deleted as well. ANOVA analysis of the agreement utilities showed that there was a statistical significance between the three treatments (p-value = 0.004). The highest average utility

Fig. 4. Example negotiation session.

Table 1. Number of agreements.

Variable	Young (Al)	Mature (Lino)	Older (Jean)
Total negotiations	35	22	18
Agreements	28	19	14
No agreements	7	3	4
Agreement rate	80%	86%	78%

was achieved by "mature" agents, followed by "older" and "younger" agents. The results of a two-tailed t-test suggest that the difference was significant between "mature" and "younger" ($p = 0.000$) and "mature" and "older" agents ($p = 0.048$). There was no statistically significant difference between the "younger" and "older" agents. Table 2 shows the utilities achieved by the agents and by the human subjects.

Table 2. Utility of agreements.

Agent type	Agent utility	Buyer utility
Young (Al)	59.00%	46.66%
Mature (Lino)	74.23%	38.40%
Older (Jean)	65.39%	44.20%

It is interesting to note that when comparing utilities of the subjects, ANOVA did not produce any significant differences between the three groups (p = 0.398). This leads to an intriguing suggestion: the subjects seemed to engage in more integrative negotiations when facing a mature counterpart. The "mature" agent facilitated not only better deals for the seller, but also acceptable ones for the buyer.

In order to measure subjective assessments of the opponent and the outcomes, a questionnaire has been administered to the subjects at the end of the experiment. The questionnaire measured two constructs, namely, satisfaction with the counterpart, and satisfaction with the outcome. Table 3 shows the items of the questionnaire. Exploratory factor analysis showed that the pattern of item loadings on the constructs was adequate. The answers were measured using seven-point Likert scales.

Table 3. Questionnaire items.

Construct	Item
Satisfaction with the counterpart	My interaction with the counterpart was positive
	I am satisfied with the counterpart
	I enjoyed working with the counterpart towards a mutually acceptable agreement
Satisfaction with the outcome	I am satisfied with the outcome of negotiations
	I feel good about the deal we have made

The comparisons of the results of the subjective measures in terms of average item responses are given in Fig. 5.

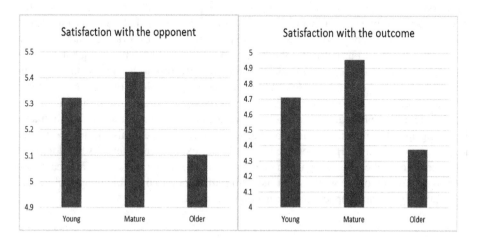

Fig. 5. Subjective assessments

ANOVA test did not reveal any significant differences in the assessments of the three categories of agents. Therefore, although one type of agents performed significantly better than the other two types, this had no statistically significant impact the subjective assessments. What is even more striking, the "mature" agent had the highest absolute scores both in terms of the satisfaction with the opponent, as well as the satisfaction with the outcomes. This is despite the fact that the subjects had the lowest average agreement utilities when negotiating with the "mature" opponent.

5   Conclusions

The current study examined the impact of a presence of a photo image in agent vs. human negotiations on negotiation outcomes. Specifically, three male images of different age were assigned to three groups of identical agents. The results of the experiments suggest that the "mature" image led to significantly superior performance in terms of utility of agreements, as compared to other images ("younger" and "older"). Moreover, the "mature" image had led to the lowest number of offers, suggesting higher efficiency of an agent. What is most interesting is that the success of this agent did not come at the expense of the lower utility of the human subjects, implying that mature image may induce more integrative negotiations.

In terms of subjective assessments, no significant differences have been detected among the groups of participants, despite the fact that there were differences in terms of utilities achieved by the agent sellers. What is even more striking, the subjects who negotiated with the "mature" agent had the lowest agreement utility levels, and yet they have had highest satisfaction levels both with the counterpart, and the agreement achieved.

A possible explanation of the results would require a further investigation into the causes of the subjects' behavior and impressions. It seems that subjects take more seriously an opponent who seems to be mature and have experience, rather than a younger individual. Furthermore, the subject might have felt that an older individual has lesser chance of keeping up with the realities of present-day products and technologies. This suggests an upside-down U-shaped relationship between the visual age cues presented to a negotiator and the utility achieved by the corresponding agents.

Admittedly, a photograph can convey and does convey more information than age, including facial features, dressing, background, etc., which can also influence a subject's decision making. However, we believe in our experimental settings the dominant characteristic of the photographic images was the age of the person. In any case, the overall conclusion of the work is that including different photographic images of people may have significant impacts on the negotiation outcomes.

References

1. Vahidov, R., Kersten, G., Saade, R.: An experimental study of software agent negotiations with humans. Decis. Support Syst. **66**, 135–145 (2014)
2. Vahidov, R., Saade, R., Yu, B.: The effects of interplay between negotiation tactics and task complexity in software agent to human negotiations. Electron. Commer. Res. Appl. **26**, 50–61 (2017)
3. Gefen, D., Straub, D.W.: Consumer trust in B2C e-Commerce and the importance of social presence: experiments in e-Products and e-Services. Omega **32**, 407–424 (2004)
4. Aldiri, K., Hobbs, D., Qahwaji, R.: The human face of e-business: engendering consumer initial trust through the use of images of sales personnel on e-commerce web sites. Int. J. E-Bus. Res. **4**, 58–78 (2008)
5. Nowak, K.L., Biocca, F.: The effect of the agency and anthropomorphism on users' sense of telepresence, copresence, and social presence in virtual environments. Presence Teleoperators Virtual Environ. **12**, 481–494 (2003)
6. Qiu, L., Benbasat, I.: An investigation into the effects of Text-To-Speech voice and 3D avatars on the perception of presence and flow of live help in electronic commerce. ACM Trans. Comput.-Hum. Interact. **12**, 329–355 (2005)
7. van 't Wout, M., Sanfey, A.G.: Friend or foe: the effect of implicit trustworthiness judgments in social decision-making. Cognition **108**, 796–803 (2008)
8. Yuasa, M., Mukawa, N.: The facial expression effect of an animated agent on the decisions taken in the negotiation game. In: CHI 2007 Human Factors in Computing Systems, San Jose, CA, USA, pp. 2795–2800. ACM (2007)
9. Citera, M., Beauregard, R., Mitsuya, T.: An experimental study of credibility in e-negotiations. Psychol. Mark. **22**, 163–179 (2005)
10. Suzuki, K., Tamada, H., Doizaki, R., Hirahara, Y., Sakamoto, M.: Women's negotiation support system—as affected by personal appearance versus use of language. In: Chung, W., Shin, C. (eds.) Advances in Affective and Pleasurable Design. AISC, vol. 483, pp. 221–230. Springer, Cham (2017). https://doi.org/10.1007/978-3-319-41661-8_22
11. de Melo, C.M., Carnevale, P., Gratch, J.: The effect of expression of anger and happiness in computer agents on negotiations with humans. In: The 10th International Conference on Autonomous Agents and Multiagent Systems-Volume 3, pp. 937–944. International Foundation for Autonomous Agents and Multiagent Systems (2011)
12. Haselhuhn, M.P., Wong, E.M., Ormiston, M.E., Inesi, M.E., Galinsky, A.D.: Negotiating face-to-face: men's facial structure predicts negotiation performance. Leadersh. Q. **25**, 835–845 (2014)

Applications of Group Decision and Negotiations

Analyzing Conflicts of Implementing High-Speed Railway Project in Central Asia Using Graph Model

Shawei He[(⊠)], Ekaterina Flegentova, and Bing Zhu

College of Economics and Management,
Nanjing University of Aeronautics and Astronautics, Nanjing 211106, China
shaweihe@nuaa.edu.cn, davinovo@126.com,
davinovo@163.com

Abstract. Conflicts arise when the proposed construction of a high-speed railway project in Central Asia affects the interests of Central Asian nations located along the route. By considering the national governments in Central Asia as decision makers, their possible actions in dealing with the conflicts are analyzed by using the graph model, a conflict resolution methodology. Three criteria, geological locations, political relations, and environmental concerns, are taken into account to accurately determine the preferences of these nations. The stabilities and equilibria of the model are calculated to provide potential strategic resolutions for these nations. The equilibrium that can take place in reality indicates that Kazakhstan and Uzbekistan can support a modified project. The opposition from Kyrgyzstan and Turkmenistan calls for appropriate resolutions from China in order to secure the successful implementation of the project.

Keywords: High-speed railway · Central asia · Graph model
Conflict resolution

1 Introduction

Central Asia is strategically important as the crossroad of the Eurasian Continent. Historically, it had facilitated the spread and interaction of civilizations via the Silk Roads, in ancient China, Persia, ancient India, and ancient Greece. According to the modern definition, Central Asia refers to the region from the Caspian Sea to China, bordering Russia to the North and Afghanistan to the South. It consists of five nations with the affix of "stan", i.e., Kazakhstan, Uzbekistan, Kyrgyzstan, Tajikistan, and Turkmenistan. With a population of around 70 million and GDP per capita below 10,000 US Dollars (UN DESA 2017), the five nations have all been experiencing industrialization since their independence from the Soviet Union in the early 1990s. Modernizing infrastructure is among the priorities of the national governments, as most of the existing infrastructure has not been renovated since the Soviet Era, which hinders growth of the economy within the region.

Railway is an important means of transportation in Central Asia. It offers large capacity, low cost, and high resilience to adverse weather conditions. The existing railway system in Central Asia was inherited from the Soviet infrastructure. The railway routes connect China and Mongolia with Europe via Kazakhstan. The current railway system in Central Asia is described as "open but blocked" in the eastern part of the region due to different gauges and "missing segments" (He 2016).

The difference in gauges between Kazakhstan and China affects the efficiency of the two countries' rail transportation. To the west, some cities have not been incorporated into the rail network: hence, they are not able to enjoy the profits and opportunities brought by communication with other regions.

As part of the "Belt and Road Initiative", a high-speed railway (HSR) project has been proposed by China in cooperation with Central Asian countries, as shown in Fig. 1. The new railway will use a gauge different from the existing 1520 mm standard to connect large cities in Central Asia (He 2016). This new system is part of a larger project building railways from Urumqi in China to Tehran in Iran. After the completion of this project, cargo as well as passengers will be transported from China to Iran, via capital cities of the four republics, which include Uzbekistan, Kyrgyzstan, Tajikistan, and Turkmenistan. This will China to share its economic achievements with the countries along the route.

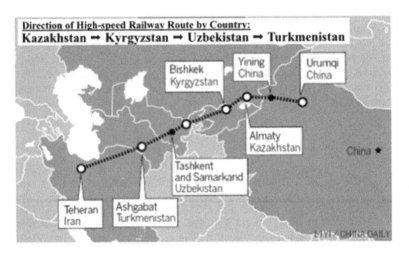

Fig. 1. Route of High-speed Rail Project in Central Asia (Modified based on http://www.chinadaily.com.cn/china/2015-11/21/content_22506412.htm)

Central Asia is well-known for its environmental fragility. The two major rivers in this region are Syr Darya and Amu Darya, flowing from the Tianshan Mountain westwards into the Aral Sea. The area through which the two rivers flow is called the Aral Sea Basin, as depicted in Fig. 2.

During the Soviet era, water was diverted from the two rivers to irrigate cotton fields in the north, however only half of the water actually reached the crops (McCray 1999). Water flow along the basins was reduced, and the Aral Sea shrank by 90% in size (Micklin and Aladin 2008). After the five nations gained independence in 1991,

Fig. 2. Countries in Aral Sea Basin (modified, based on Nandalal and Hipel (2007))

the Aral Sea Basin became a transboundary area overnight. Disputes arose over the allocation of water resources, sometimes intensified by ethnic diversity, bureaucracy, and ill-functioning infrastructure.

Environmental impacts should be considered in constructing the high-speed railway (HSR) in Central Asia. Compared with other means of transportation, such as private cars and buses, HSR will produce less carbon emissions per capita. However, the construction of HSR can still affect the environment in the following ways:

- It may deteriorate wetlands and affect water quality.
- It may occupy agricultural land, which is precious in Central Asia.
- After the completion of the project, the increase in pollution along the railway, caused by either passengers or new inhabitants, will result in higher demand for water, thereby exerting pressure on the limited water resources.
- The noise and vibrations along the railway may also have negative effects on local residents and wildlife.

Accordingly, decision makers in these nations should evaluate the project by considering these consequences.

Potential environmental conflicts for implementing the project may arise among parties situated along the proposed railway and the two rivers. Some stakeholders welcome the project due to the potential economic benefits, while opposition may come from stakeholders with environmental concerns. From the perspective of the five national governments, the construction of HSR can improve the economy in the region. However, their attitudes towards the project depend on various factors, such as geological features along the construction route, the flow of the two rivers, and political relationships with China and neighboring countries. As the biggest economy in the region, Kazakhstan has shown an interest in building HSR with China (Tabyldy 2017). Although China has sponsored Uzbekistan and Turkmenistan for linking the conventional railways with Kazakhstan in 2014 (Arina 2016), the difficult bilateral relations between Tajikistan and Uzbekistan, and the negative view of Kyrgyzstan on railway projects passing through its territory may affect the implementation of HSR (Savi and Peremen 2017). Hence, the actions of the national governments as stakeholders towards the implantation of the HSR project proposed by China should be systematically analyzed by considering geological, political, and environmental factors. This paper raises the following related questions:

- Assuming that each national government strives to achieve maximum benefits, how can one determine the preference of each decision maker (DM) when their value systems are hard to evaluate due to difficulties in accessing information or data?
- Under different behavioral patterns, what are DMs' potential actions, or strategic resolutions, in dealing with the conflicts caused by the implementation of the HSR project?
- What can each national government learn from this conflict to guide their actions when the strategies of its own and others can be obtained?

Note that a DM in this paper refers to a stakeholder: not only can the stakeholder's interests be affected in/by a conflict, but the stakeholder can also impact the conflict by taking actions. To formally investigate these questions, an appropriate conflict analysis methodology should be employed.

2   Literature Review

Various conflict analysis methodologies have been utilized to systematically study strategic conflicts. Game theoretic models (Von Neumann and Morgenstern 1944) have been widely used to handle conflicts with multiple DMs and objectives. In most game theoretic models, the value systems for DMs are characterized by numbers, either certain or with uncertainty. In reality, payoffs in cardinal numbers such as utility values are hard to determine. In many cases, resolution of a conflict can still be obtained without the requirement of cardinal utilities. Thus, non-quantitative models are developed by using relative preferences to represent the payoffs of DMs, such as Metagame Analysis (Howard 1971) and Conflict Analysis (Fraser and Hipel 1979, 1984), in which the modeling structures are more flexible by allowing irreversible moves and introducing more solution concepts to describe the behavior patterns of DMs. The genealogy of conflict analysis methodologies is shown in Fig. 3.

The Graph Model for Conflict Resolution (GMCR) (Fang et al. 1993; Kilgour and Hipel 2005) is an extension of Conflict Analysis representing outcomes of a strategic conflict, usually called states, and moves between states – as transitions; they are represented with graphs. The behavior of DMs is analyzed under four major solution concepts: Nash stability (R) (Nash 1950, 1951), sequential stability (SEQ) (Fraser and Hipel 1979, 1984), general metarationality (GMR) (Howard 1971), and symmetric metarationality (SMR) (Howard 1971). The strategic resolutions for DMs can be obtained by decision support systems (Fang et al. 2003a, b; Kinsara et al. 2015).

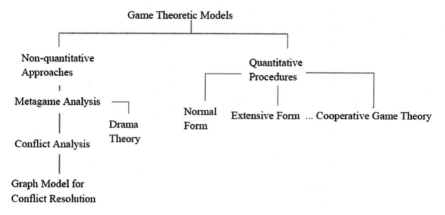

Fig. 3. Genealogy of conflict analysis methodologies (Hipel and Fang 2005)

Preferences in decision making have been extensively studied. Techniques have been utilized to assess preferences of DMs according to multiple criteria (Belton and Stewart 2002). Some factors affecting the preferences of DMs are conflicting (Keeny et al. 1994). Scoring systems like PROMETHEE (Brans and Mareschal 2005) and AHP (Analytic Hierarchical Process) (Saaty 2001) were developed to describe the value systems of decision makers by considering various criteria. Uncertainties in preferences have been studied by using fuzzy logic (Grabisch and Labreuche 2005) and grey theory (Liu and Lin 2010).

Within the paradigm of GMCR, Ke (2008) developed a GMCR model by and augment it with an AHP model used to elicit relative preferences. The AHP model considers criteria for selecting options. Multiple objectives of DMs are analyzed by Bristow et al. (2012) within the paradigm of GMCR. Objectives of DMs are compared pairwise and preferences are represented separately by different value systems. Option Prioritization approach (Fang et al. 2003a, b) has been developed to effectively represent the preferences by the options of decision makers, because the number of options in a graph model is significantly smaller than the number of outcomes. The criteria that affect the preferences should be considered so that they may be expressed more precisely.

In the remainder of the paper, the theoretical structure of the existing methodology, GMCR, is briefly introduced in Sect. 3. The two steps of investigating the HSR construction conflict, modeling and analysis, are mentioned in Sects. 4 and 5, respectively. The conclusions and further study are given in Sect. 6.

3 Graph Model for Conflict Resolution

A strategic conflict can be studied using graph model in two steps: modelling and analysis, as shown in Fig. 4. A graph model can be represented by a 4-tuple set $G = \{N, S, A, \succsim\}$ consisting of the sets of DMs (N), states (S), unilateral moves (A), and preference relations on $S\ (\succsim)$.

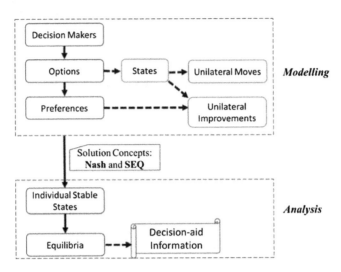

Fig. 4. Process of investigating strategic conflicts using Graph Model for Conflict Resolution (modified based on Fang et al. (1993))

Within the GMCR paradigm, DMs in a real world conflict are identified in the modelling part. The options and preferences of each DM can be obtained according to the background of the conflict. An option is a possible choice that can be taken by a DM. By knowing the options, each state in the conflict can be represented as a combination of the selection of options from all DMs.

A DM may move unilaterally from one state to another by changing its option selections. For each DM $i \in N$, the set of unilateral moves is expressed as $A_i \subseteq S \times S$. For two states $s, s' \in S$, if i has a unilateral move (UM) from s to s' in one step, this move can be denoted as $(s, s') \in A_i$.

In graph model G, each DM's preferences on S are completely determined by the relation $\succsim$ which is assumed to be complete and reflexive. In particular, $s \succ_i s'$, $s \sim_i s'$, and $s \prec_i s'$ indicate that s is more, equally, and less preferred to s' for i, respectively. The three relations can also be combined. For instance, $s \precsim_i s'$ represents that s no better than s' for i. Some UMs for a DM are also called unilateral improvements (UIs) if these UMs can result in more preferred states for the DM. Specifically, a UM for i, $(s, s') \in A_i$, is UI if $s \prec_i s'$.

In the analysis part, stabilities of states are calculated. Stabilities, also called solution concepts, are utilized to describe possible behavior patterns of DMs in real world conflicts, varying by DM's scope of moves and the perceptions on counteractions from other DMs. Four types of stabilities are investigated, including Nash stability (Nash), sequential stability (SEQ), general metarationality (GMR), and symmetric metarationality (SMR). (this was mentioned earlier) For simplicity, the two representative stabilities, Nash and SEQ, are analyzed in this paper. Theoretically, Nash and SEQ are stronger stabilities than GMR and SMR because only the favorable actions of DMs within two steps are considered. In comparison, the unfavorable moves of DMs as counteractions are included in GMR; SMR involves moves of DMs in three steps.

Thus, Nash and SEQ reflect behavioral patterns of DMs that take place in reality more commonly.

A state is Nash stable for a given DM if and only if there is no UI for the DM at this state. The mathematical definition is written as (Nash 1950, 1951): $s \in S$ is Nash stable for $i \in N$ if and only if $R_i^+(s) = \emptyset$, where $R_i^+(s)$ denotes the set of UIs at s for i.

An SEQ state for a focal DM reflects a situation at which the DM can be worse off by the subsequent sanctions from other DMs against its UIs. Theoretically, SEQ (Fraser and Hipel 1979, 1984) is defined as: $s \in S$ is SEQ for $i \in N$ if and only if for every $q \in R_i^+(s)$, there exists $r \in R_{N-i}^+(q)$, such that $s \prec_i r$, where $R_{N-i}^+(q)$ represents the set of UIs for all DMs except i, marked as $N - i$, at state q.

A state can be stable by either Nash or SEQ for a given DM. A stable state for the DM indicates an outcome of a conflict at which the given DM is unlikely to move away. A state is stable for all DMs, either Nash or SEQ, suggests an outcome that is likely to happen or useful resolution for all DMs, as no DM is inclined to move away from this state. This stable state is called equilibrium. The equilibria in a conflict are a useful output, indicating guidance of actions for DMs to follow in reality.

4 Conflict Modeling

4.1 Decision Makers and Options

In a strategic conflict, DMs are parties or groups who are concerned about their interests and are able to impact the conflict by taking actions. In the HSR disputes, four national governments in Central Asia, consisting of Kazakhstan (KZ), Uzbekistan (UZ), Kyrgyzstan (KY), and Turkmenistan (TK), are considered DMs as they are situated along the planned railway route. Tajikistan is not a DM as it is not included in the current plan. The Chinese national government (CN) is another DM, because the construction of the HSR project is of its great interest. Although Tehran is the destination of the project, Iranian national government is not included: the impacts of the HSR project in Central Asia are out of its scope of considerations. Thus, the five DMs in the conflict are KZ, KY, UZ, TK, and CN.

Each DM has at least one option in the conflict. Each central Asian nation can either agree with CN or show opposition, which can be combined into one option as "Agree": the opposition is expressed as the negation of "Agree". If KZ agrees with CN, for instance, its option "Agree" is marked with a "Y". Otherwise, an "N" is labeled with this option. CN can initiate the construct plan or suspend the construction. Considering the possible oppositions from the four central Asian governments, CN could also modify the project, by providing financial support to these nations, changing the detailed construction plan, or transferring green technologies to alleviate possible damage to the environment along the route. The options for CN are written as "Initiate" and "Modify". The negation of "Initiate" refers to the suspension of the project; the opposite of "Modify" means no change to the original plan. The selection of the two options for CN implies different meanings. For example, Y for "Initiate" and N for "Modify" means that CN will implement the original project; N for "Initiate" and Y for "Modify" denote that CN suspends the project although it is modified. The DMs and

Table 1. Decision makers, options, and sample states

Decision makers	Options	Sample state
KZ	(1) Agree	N
KY	(2) Agree	N
UZ	(3) Agree	N
TK	(4) Agree	N
CN	(5) Initiate	Y
	(6) Modify	Y

their options are listed in Table 1. The options of all DMs are assigned a number followed by a half parenthesis, for labeling purposes.

As defined in Sect. 3, a state, regarded as a possible scenario of conflict, is a combination of option selections for all DMs. As each option can be chosen by the corresponding DM or not, there are 2^6 states for the total of 6 options. A sample state is also listed in Table 1, indicating the scenario in which CN's implementation of modified project is opposed by all of the central Asian nations along the route.

4.2   Multiple Criteria Preferences

The preferences of the DMs in the HSR project conflict are determined from the perspectives of the nations' geological locations, political relationships, and environmental concerns. Geological location of a nation refers to its position along the HSR. According to Fig. 1, countries are linked in a sequence starting from CN in the east westbound to TK via KZ, KY, and UZ. A nation to the west relies on the connectivity of the route in its eastern neighbors. Hence, the rear nations along the route are more dependent on the actions of the nations in the front. Political relations of a nation with others is another criterion to consider.

According to the background scanning of the HSR project, UZ and TK are likely to favor the project as they had similar collaboration with CN in the past (Savi and Peremen 2017). KZ could also be supportive as it has shown an interest to the HSR project (Tabyldy 2017). KY could be less favorable because it reportedly dislikes the connection of its railway to UZ due to their territorial disputes. Environmental factors are also important in shaping the preferences of the DMs. By taking into account possible impact on water resources and soil, the position of each nation along Syr Darya and Amu Darya matters. A downstream DM is more concerned with the environmental impacts as it is more vulnerable to potential environmental damages than an upstream DM. KZ, UZ, and TK are at the relatively downstream position compared with KY. The rules to determine the preferences according to the three criteria are listed in Table 2.

Table 2. Preference rules by criterion

Geological	Political	Environmental
CN - KZ - KY - UZ - TK	Favor: UZ, TK, KZ Dislike: KY	Syr Darya: KY - UZ - KZ Amu Darya: UZ - TK

Preferences are often represented in terms of the ranking of states in a conflict. When the number of states is large, Option Prioritization Approach (Fang et al. 2003a, b) is employed by ranking the options for all DMs instead of the states, because the number of options is much smaller than that of the states. By using this approach, the preferences for a DM are expressed by the statements of options connected by logical symbols, such as AND (&), OR (|), NOT (-), IF, and IFF meaning if and only if. These statements are ranked from the most important for the focal DM at the top to the least important at the bottom. The preferences for DMs in the HSR conflict are investigated by the three above mentioned criteria: geological position, political relations, and environmental concerns.

Table 3. Preference statements for DMs by criterion

Geological Position		Political Relations		Environmental Concerns	
Ranking	2nd	1st		3rd	
CN 1 2 3 4	Support along the route by vicinity.	1&2&3&4 6 IF (-1|-2|-3|-4)	All support Modify if at least one opposition	6 IF -2	Modify if KY opposes
5 IF 1 (5&6) IF -1	At least KZ supports				
Ranking	3rd	1st		2nd	
KZ 2 IF 1	Connection via KY if KZ agrees	1 IF 6 5 6 IF -1	KZ wishes CN to modify	(-3)IF 1 (-2)&(-3) IF -1 (-2)|(-3) IF -1 (-4) IF -1	Opposition from upstream UZ Coalition from other nations if opposes
Ranking 2nd		1st		3rd	
KY -3	Expect opposition from west neighbor	-5 6	No original project Better oppose		
2 IF 1	Support if east neighbor supports	-2			
Ranking 2nd		1st		3rd	
UZ 1&2	Connection via KZ and KY	3 IF 6	Support the modified project	6 IF (-1) & 5 6 IF (-2) & 5	Hope modify if downstream and upstream oppose original
2 IF 3	Connection via KY				
Ranking 1st		3rd		2nd	
TK 1&2&3IF 4	Connection via precedent nations	6 4 IF 6	Support the modified	(-3) IF 5 (-4) IF 3	Upstream opposes o Oppose if upstream supports

First, preference statements are written for each DM using each criterion. For example, when considering the political relations, the most desired outcome is the support from all other nations, denoted as the selection of option 1, 2, 3, and 4. Thus, the first preference statement for CN is written as 1&2&3&4, positioned at the top in the corresponding cell of Table 3. To follow up, CN will modify the project (Option 6) if at least one other nation shows opposition, which can be expressed as 6 IF (-1|-2|-3|-4) placed below the first statement in Table 3. Note that we assume the three criteria to be independent. Thus, the preference statements are elicited by considering only one criterion at one time.

For each DM, the importance of these criteria is different, represented by a ranking from 1 to 3. For instance, geological position is the most important criterion for TK: as it is situated at the rear of the route, TK is more dependent on the connectivity of HSR than other nations. The environmental concerns are important at the second place, as TK is a downstream country along the Amu Darya. The preference statements for each DM by criterion are shown in Table 3. Brief explanations to these statements are provided to the right. The complete preference statements for the DMs are listed in Table 4, reflecting not only the ranking of the statements within a criterion but also the ranking of the criteria for each DM.

Table 4. Complete preference statements for DMs

CN	1&2&3&4	Political
	6 IF (-1|-2|-3|-4)	
	1	Geological
	2	
	3	
	4	
	5 IF 1	
	(5&6) IF -1	
	6 IF -2	Environmental
KZ	1 IF 6	Political
	5	
	6 IF -1	
	(-3)IF 1	Environmental
	(-2)&(-3) IF (-1)	
	(-2)|(-3) IF (-1)	
	(-4) IF -1	
	2 IF 1	Geological
KY	-5	Political
	6	
	-2	
	-3	Geological
	2 IF 1	
UZ	3 IF 6	Political
	1&2	Geological
	2 IF 3	
	6 IF (-1) & 5	Environmental
	6 IF (-2) & 5	

(continued)

Table 4. (*continued*)

TK	(1&2&3) IF 4	Geological
	(-3) IF 5	Environmental
	(-4) IF 3	
	6	Political
	4 IF 6	

5   Stability Analysis Using GMCR II

Individual stabilities and equilibria for DMs in the HSR project conflict are analyzed using a decision support system GMCR II (Peng et al. 1997). GMCR II was designed to carry out calculations for stabilities by typing information about DMs, options, and preference statements in the modeling subsystem.

A list of feasible states, the transition of states, and the state rankings for each DM can be displayed. Stabilities for DMs and equilibria can be obtained as the output of the system. In-depth analysis can also be implemented such as sensitivity analysis to examine whether and to what extent the change in the model output is affected by the input information. The structure of GMCR II is described in part (I) of Fig. 5.

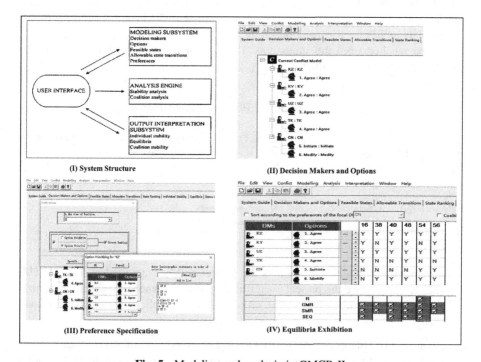

Fig. 5. Modeling and analysis in GMCR II

To start modelling using GMCR II, the five DMs and their options are typed into the panel called "DMs and Options" as shown in part (II) of Fig. 5. The total of 2^6 states can be displayed in the panel called "feasible states", each of which is assigned a number ranging from 1 to 64. The preference statements can be specified in the "Preferences" panel by selecting "Option Prioritization", as displayed in part (III) of Fig. 5. By clicking the "Analysis" on the top bar in part (IV) of Fig. 5, the equilibria of HSR construction dispute under the four solution concepts can be displayed. In this paper, the two representative solution concepts, Nash and SEQ, are analyzed: they are "stronger" than GMR and SMR, and therefore can suggest outcomes that are more likely to happen and provide more meaningful resolutions. According to the screen exhibiting the equilibria in Fig. 5, state 54 is Nash equilibrium and the SEQ equilibria are states 16, 40, and 48.

An evolution analysis is further carried out to examine which equilibrium can be reached from the starting state by initiating UIs from DMs, also called the status quo. This particular equilibrium can be used to interpret the outcome of the HSR construction conflict. As an assumption in GMCR, the order of UIs does not affect the final outcome. The path of the evolution is demonstrated by Table 5. The status quo, state 1, represents the scenario before the implementation of the HSR project, at which point no other nation agrees with this project. Starting from state 1, CN can initiate a UI by adapting to a modified plan, resulting in state 49. To follow up, KZ's UI by changing option (1) from N to Y leads to state 50. The evolution of the conflict stops at state 54, as no DM has UI at this state after it evolved from state 50 by UZ's subsequent UI. Further analysis has been carried out, indicating that state 54 is the only equilibrium that can be reached from the status quo. Thus, Nash equilibrium is regarded as the possible outcome of the conflict.

Table 5. Evolution of HSR construction conflict

			State 1 (Status Quo)	State 49	State 50	State 54
KZ	1)	Agree	N	N ⟶	Y	Y
KY	2)	Agree	N	N	N	N
UZ	3)	Agree	N	N	N ⟶	Y
TK	4)	Agree	N	N	N	N
CN	5)	Initiate	N ⟶	Y	Y	Y
	6)	Modify	N ⟶	Y	Y	Y

At state 54, the modified version of the project will be supported by KZ and UZ. KY and TK will oppose the project even if it is modified. Hence, not all of the Central Asian countries will support the HSR project. Several implications can be obtained from state 54:

(1) As state 54 is the equilibrium at which the HSR project can receive the support from the most Central Asian nations, the support from all the nations cannot be realized regardless of the effort from CN under the current preference settings.

(2) According to Table 3, TK will oppose the project although it prefers to support the modified version from the viewpoint of political relations. TK's opposition is due to its comprehensive consideration of all criteria including the geological location and environmental concerns. To gain support from TK, CN should mitigate potential environmental damages in TK by, for example, transferring clean technology and by allocating special funding for compensating the damages.

(3) KY will also oppose the project out of the political and geological concerns. China should be proactive in holding multilateral negotiations with UZ and KZ to solve the territorial disputes related with the project.

(4) Comprehensive initiatives should be taken by CN to attract the Central Asian nations in implementing the HSR project with CN. Some details of the construction plan may be redesigned to protect the environment in the four nations. Financing solution scan be adopted. For example, special funding for environmental protection and mortgages to relieve the financial burden of the construction can be provided.

6   Conclusions and Further Study

In this paper, conflicts caused by the construction of high-speed railway project in Central Asia from the perspectives of geological locations, political relations, and environmental concerns are analyzed using Graph Model for Conflict Resolution. A multiple criteria preference structure under the framework of Option Prioritization is designed. This new preference structure can describe the preferences for DMs in the conflict more precisely by considering the impacts of the three criteria on the preferences. The equilibrium obtained by GMCR II indicates that the national governments of Kazakhstan and Uzbekistan will support the project when it is modified. China should seek support from Kyrgyzstan and Turkmenistan using various means, including multilateral negotiation and financial aid.

The limitation of the model presented in this paper is inadequate specificity; the options for DMs need to be elaborated further. For example, modification as option (6) includes financial support, change of the project plan, and transfer of technologies. Further study should be carried out to analyze this conflict with more specific options. Besides, machine learning techniques can be applied to the determination of preferences in order to improve the accuracy of describing preferences. As the current HSR project is still under planning, the evolution of equilibria at different stages of the conflict can be studied by considering the time frame. Moreover, the conflict model can be expanded by taking into account the influence of global powers such as Russia, the United States, and the European Union.

References

Arina, M.: Chinese Silk Road to be Held by Railways of Central Asia (In Russian). Russian Institute of Strategic Studies (2016). https://riss.ru/analitycs/27356/. Last accessed 20 Dec 2017

Belton, V., Stewart, T.: Multiple criteria decision analysis: an integrated approach. International **142**(6), 192–202 (2002)

Brans, J.P., Mareschal, B.: Promethee methods. In: Greco, S. (ed.) Multiple Criteria Decision Analysis: State of the Art Surveys. ISOR, vol. 78, pp. 163–186. Springer, New York (2005). https://doi.org/10.1007/0-387-23081-5_5

Bristow, M., Hipel, K., Fang, L.: Ordinal preferences construction for multiple-objective multiple-participant conflicts. In: IEEE International Conference on Systems, Man, and Cybernetics, pp. 2418–2423. IEEE (2012)

Fang, L., Hipel, K., Kilgour, M.: Interactive Decision Making: The Graph Model for Conflict Resolution, vol. 3. Wiley, New York (1993)

Fang, L., Hipel, K., Kilgour, M., Peng, X.: A decision support system for interactive decision making-Part I: model formulation. IEEE Trans. Syst. Man Cybern. Part C Appl. Rev. **33**(1), 42–55 (2003a)

Fang, L., Hipel, K., Kilgour, M., Peng, X.: A decision support system for interactive decision making - part II: analysis and output interpretation. IEEE Trans. Syst. Man Cybern. Part C Appl. Rev. **33**(1), 56–66 (2003b)

Fraser, N., Hipel, K.: Solving complex conflicts. IEEE Trans. Syst. Man Cybern. **9**(12), 805–816 (1979)

Fraser, N., Hipel, K.: Conflict Analysis: Models and Resolutions. Series 9, vol. 11. North-Holland, New York (1984)

Grabisch, M., Labreuche, C.: Fuzzy measures and integrals in MCDA. In: Multiple Criteria Decision Analysis: State of the Art Surveys. International Series in Operations Research & Management Science, vol. 78, pp. 563–604. Springer, New York (2005). https://doi.org/10.1007/0-387-23081-5_14

He, H.: Key challenges and countermeasures with railway accessibility along the Silk Road. Engineering **2**(3), 288–291 (2016)

Hipel, K., Fang, L.: Multiple participant decision making in societal and technological systems. In: Systems and Human Science, for Safety, Security, and Dependability: Selected Papers of the 1st International Symposium SSR 2003, Osaka, Japan, November 2003, p. 1. Elsevier (2005)

Howard, N.: Paradoxes of Rationality: Theory of Metagames and Political Behavior. MIT Press, Cambridge (1971)

Ke, Y.: Preference elicitation in the graph model for conflict resolution. Master's thesis, University of Waterloo (2008)

Keeney, R., Raiffa, H., Rajala, D.: Decisions with multiple objectives: preferences and value trade-offs. J. Oper. Res. Soc. **45**(9), 1093–1094 (1994)

Kilgour, M., Hipel, K.: The graph model for conflict resolution: past, present, and future. Group Decis. Negot. **14**(6), 441–460 (2005)

Liu, S., Lin, Y.: Introduction to grey systems theory. Underst. Complex Syst. **68**(2), 1–18 (2010)

Mccray, T.R.: Enviro-economic imperatives and agricultural production in Uzbekistan: modern responses to emergent water management problems. Dissertation Abstracts Int. **59** (1999)

Micklin, P., Aladin, N.: Reclaiming the Aral Sea. Sci. Am. **298**(4), 64–71 (2008)

Nandalal, K., Hipel, K.: Strategic decision support for resolving conflict over water sharing among countries along the Syr Darya River in the Aral Sea Basin. J. Water Resour. Plan. Manage. **133**(4), 289–299 (2007)

Nash, J.: Equilibrium points in n-person games. Proc. Nat. Acad. Sci. USA **36**(1), 48–49 (1950)

Nash, J.: Non-cooperative games. Ann. Math. **54**(2), 286–295 (1951)

Neumann, V., Morgenstern, O.: Theory of Games and Economic Behavior, 1st edn. Princeton University Press, Princeton (1944)

Kinsara, R., Petersons, O., Hipel, K., Kilgour, M.: Advanced decision support for the graph model for conflict resolution. J. Decis. Syst. **24**(2), 117–145 (2015)

Saaty, T.L.: Analytic Hierarchy Process. Encyclopedia of Biostatistics. Wiley, New York (2001)

Savi, K., Peremen, M.: Trade Development in the CAREC Region: the Potential of Central Asian Railways (In Russian) (2017). http://mirperemen.net/2017/05/razvitie-torgovli-v-regione-cares-potencial-zheleznyx-dorog-centralnoj-azii/. Last accessed 12 Dec 2017

Tabyldy, K.: Where Will the Silk Road Lead? (In Russian), Sputnik (2017). https://ru.sputnik.kg/analytics/20170116/1031299276/kuda-privedet-shelkovyj-put.html. Last accessed 12 Dec 2017

UN DESA: World population prospects, the 2017 Revision, Volume I: comprehensive tables. New York United Nations Department of Economic & Social Affairs (2017)

Strategic Negotiation for Resolving Infrastructure Development Disputes in the Belt and Road Initiative

Waqas Ahmed$^{(\boxtimes)}$, Qingmei Tan, and Sharafat Ali

College of Economics and Management, Nanjing University of Aeronautics
and Astronautics, Nanjing, Jiangsu, China
waqas19@nuaa.edu.cn

Abstract. Regional economic corridors are playing a role in uplifting the infrastructure of developing countries. But, such integrations are prone to some challenges emerging from the multilevel system of governance in participating countries. It is necessary that legitimate stakeholders get involved at national, provincial and local levels using collaborative planning and development. Exclusion at any level would ultimately lead to unsolicited and undesirable outcomes. The present study uses Graph Model for Conflict Resolution (GMCR) as a primary conflict resolution tool to resolve Pakistan Railway (PR) infrastructure development disputes under the China-Pakistan Economic Corridor (CPEC). This tool takes into consideration interests of all stakeholders. It could be used for future planning by policymakers.

Keywords: Regional infrastructure · Planning · Conflict analysis
Belt and road initiative · China-Pakistan economic corridor

1 Introduction

The development of economic corridors, in the regions of strategic importance, not only benefits the interconnected economies but also opens the avenues for their economic prosperity. Enhanced connectivity by infrastructure and communication leads to international cooperation including bilateral and multilateral engagements [1]. The cross-border regionalism involves a multi-level system of governance ranging from international to national, provincial, and local levels coupled with stakeholders with either economic or political objectives. Planning of regional integration through a corridor is a multitier process, prone to an array of regulatory, coordination and investment challenges.

Many of these seemingly practical and technical viable projects overlay with political and institutional challenges at domestic levels [2, 3]. It is crucial to involve all stakeholders in the planning process [3–6] as it assures the wide credence of the development of the plan [6]. The stakeholders may have conflicting objectives. Moreover, they may have common objectives but may have their conflicting strategies to attain common objectives [7]. However, dialogue, cooperation, and collaboration among various stakeholders in achieving a common goal play a vital role in preserving

© Springer International Publishing AG, part of Springer Nature 2018
Y. Chen et al. (Eds.): GDN 2018, LNBIP 315, pp. 154–166, 2018.
https://doi.org/10.1007/978-3-319-92874-6_12

vested interests and in providing a win-win solution for all [3, 8]. Failure to do so may make underprivileged and aggrieved regions feel neglected. In addition to this, the lack of shared vision [9], unavailability of information, lack of trust, political motives and lack of coordination among governmental departments and institutions [5] may lead to a serious conflict hampering the execution of the project as agreed [10].

The Belt and Road Initiative (BRI), an impetus to the doctrine of "constructive engagement" [11], is also exposed to several challenges. This initiative is multitier with numerous infrastructural and developmental projects, articulating the vision of connecting regions through numerous trade corridors. The CPEC is a flagship project under the BRI. The execution of many projects under CPEC created a controversy though it has a clear vision. The authors considered the case study of controversy regarding the upgrading and route selection of Railway corridors in Pakistan. There is a growing need to provide the basis for strategic planning for such development projects and this research provides the foundations for such planning considering provincial governments, the federal government, and Chinese stakes. A suitable conflict resolution strategy assures the best possible solution to the conflict. This could help prompt execution of the projects without hampering investors' risk orientation and international credibility of the country regarding ease of business.

The study aims to provide the basis to find out the most appropriate solution using a scientific approach acceptable to all legitimate stakeholders. This research could be a pioneer study using decision-analysis in regional integration projects of South Asia. A systematic analysis of the conflict could provide a better understanding of conflict emergence and could also provide options for its avoidance. This study sets a benchmark for coming up with a win-win solution considering explicit and implied interests of all stakeholders, not only in the current conflict but in all conflicts in other similar projects. The present research develops a formal conflict model based on the graph model for conflict resolution (GMCR) [7, 12]. This decision analysis technique is very suitable under the circumstances as it requires very little information regarding a conflict in comparison to, for example, game theory and uses the available information in a systematic and scientific way to find reasonable and feasible solutions of a conflict [12, 13]. The study traces out some reasonable solutions to the conflict following the conflict analysis strategy suggested in [12].

2 Background of the Conflict

Upgrading and restructuring of PR got government's attention in 2006 when a plan to extend railway linking Havelian to Kashgar – China was proposed in Musharraf era [14]. In 2006, during the Chinese President's visit to Pakistan, Chinese enterprises took interest in the interconnection and construction of Gwadar as a win-win scenario for the energy security of Pakistan and China. In 2008, China and Pakistan signed the cooperation documents and a framework for cooperation [15]. In 2013, both countries inked an MOU for CPEC framework and established a Joint Cooperation Committee (JCC) for the CPEC that initiated the core planning at the federal level [16] without considering the provinces. Based on such dialogues at the federal level, 12 CPEC projects were given priority as EHPs, along with numerous agreements and MoUs worth US$28 billion during Chinese president's visit to Pakistan in 2015.

The upgrading of the ML-I (Eastern route) (See Fig. 1) was given priority as EHP. But the route selection of railway corridor as EHP attracted criticism from Baluchistan and KPK as this route is in the eastern part of Pakistan. However, there was no project related to infrastructure development or upgrade in these provinces as EHPs [17–19] which resulted in the public uproar and loss of trust as the distribution of projects in KPK and Baluchistan was deemed unfair [20]. Moreover, asymmetric information and lack of collaborative planning led to serious conflict between the western provinces and the GOP. There were four railway corridors planned in the PR strategic plan [21]. The national parties of western provinces blamed GOP for changing original route of CPEC, claiming that such change would undermine the interests of the Baloch and the Pashtuns. The federal government rejected the allegation of exclusion [22]. Later, Chief Minister of KPK threatened to withdraw his cooperation in the acquisition of land for the CPEC projects [23]. Federal Government tried to get political consensus organizing two All Parties Conferences (APCs) on May 2015 [24] and January 2016 [25].

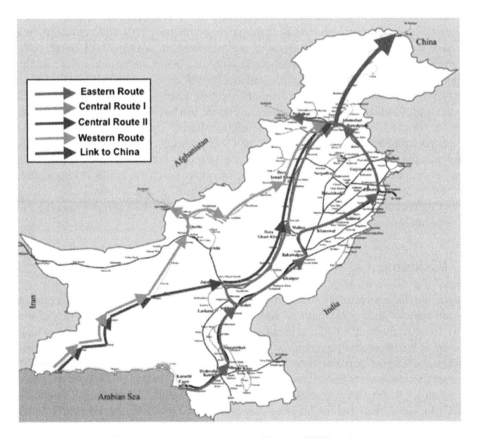

Fig. 1. Suggested railway corridors in CPEC project

However, because of asymmetric illustration of vision and lack of trust that GOP woud develop infrastructure [26] gave the impression that the GOP was playing tricks [27]. The Pakistani premier reassured the involved parties that there was no change of route of the CPEC, the western route would be constructed first along with railway, fiber optic cable and other infrastructure by July 2018 [25]. This dilemma led to China's interference in the dispute. The Chinese expressed their concerns [28] and made a reference to the Monographic Study on Transport Planning of CPEC [29]. The distrust and dispute still exist and this paper attempts to come up with the best possible solutions under the given scenario.

3 Conflict Analysis Approach

The GMCR is a simple and flexible approach designed for conflict analysis [12, 30]. It has been used in a wide range of areas such as military strategies, peace-keeping activities, environmental management, natural resources and water resource issues, urban planning to name a few. It is a suitable technique to analyze the conflict emerging during the planning and implementation of the CPEC project under the BRI as it puts the complex strategic decisions into perspective and provides better understating of the nature of the conflict thereby envisioning the potential solutions as well [12, 31]. This section provides a brief introduction to the graph model and definitional concepts used for the analysis in the GMCR approach.

Definition 1 (Graph Model for Conflict Resolution): Fang et al. [12] present the definitions for conflict models within the framework of the GMCR with the N-DMs, N $\geq$ 2. Having the set of all states $X = \{x_1, x_2, \ldots, x_k\}$, the preference structure for DM i, $\{\psi_i, \sim_i\}$, where $x_1 \succ_i x_2$ and $x_3 \sim x_4$ implying that state x_1 is preferred to x_2 and x_3 and x_4 are equally preferred by the DM i. Having the set of DMs N and the set of all states X, the set of states that are preferred, the set of states equally preferred and the set of less preferred states to x for DM i is $\Psi_i^+(x) = \{x_m : x_m \succ_i x\}$ $\Psi_i^=(x) = \{x_m : x_m \sim_i x\}$ and $\Psi_i^-(x) = \{x_m : x_m \succ_i x_m\}$, respectively.

Definition 2 (Reachable list & Unilateral Improvement (UI)): The reachable list [12], in a GMCR model for DM i from state $x \in X$ symbolized as $R_i(x) \subseteq X$, is the set of states that a DM could move to from state x. It can be subdivided as the set of UI from a state x, $R_i^+(x) = R_i(s) \cap \Psi_i^+(x)$, the set of equally preferred independently reachable states from x, $R_i^=(x) = R_i(x) \cap \Psi_i^=(x)$, and the set of unilateral improvements from state x, $R_i^-(x) = R_i(x) \cap \Psi_i^-(x)$.

The solution concepts used in the present conflict analysis are defined as:

Definition 3 (Nash Stability) (Nash): A state $x \in X$ is Nash stable (Nash) for DM i, denoted by $x \in X_i^{Nash}$, if and only if (IFF) $R_i^+(x) = \phi$ [32, 33]. In this case, there is no unilateral improvement for DM i from the state x.

Definition 4 (General Meta-rationality): State $x \in X$, in N-DMs conflict, is general meta-rational (GMR) for DM i, symbolized as $x \in X_i^{GMR}$, IFF $\Box$ $x_1 \in R_i^+(x)$ there is $x_2 \in R_j(x_1)$ such that $x_2 \in \Psi_i^-(x) \cup \Psi_i^=(x)$ [34]. It implies that state x is GMR for DM i

IFF DM i is sanctioned to move from this state by opponent DM j by subsequently moving to other state x_2 which is less preferred to initial state for DM i.

Definition 5 (Symmetric Metrationality): A state $x \in X$, in N-DMs conflict, is symmetric meta-rational (SMR) for DM i, signified as $x \in X_i^{SMR}$, IFF $\square$ $x_1 \in R_i^+(x)$ there is $x_2 \in R_j(x_1)$ such that $x_2 \in \Psi_i^=(x) \cup \Psi_i^-(x)$ and $x_3 \in \Psi_i^=(x) \cup \Psi_i^-(x)$, $\square$ $x_3 \in R_i(x_2)$ [34]. It implies that a state x is SMR for a DM i IFF any UI from x to x1 for DM i could be sanctioned by opponent DM j by moving to other state x2 and DM i is unable to escape this sanction. In this situation, DM i prefers to stay at state x.

Definition 6 (Sequential Stability): A state $x \in X$, in a N-DMs conflict, is sequentially stable (SEQ) for DM i, indicated as $x \in X_i^{SEQ}$, IFF $\square$ $x_1 \in R_i^+(x)$ there is $x_2 \in R_j(x_1)$ such that $x_2 \in \Psi_i^=(x) \cup \Psi_i^-(x)$ [7]. Simply, a state x is SEQ for DM i IFF DM i's every UI from x is sanctioned by a UI of DM j (credible sanction).

The stability of the states for each DM is analyzed under the stability concepts of Nash, GMR, SMR and SEQ in the framework of the GMCR approach. The states satisfying the stability condition under a particular stability concept for all DMs in a conflict is deemed to be equilibriums of the conflict. An equilibrium is strong if it qualifies all stability definitions [12, 13, 35].

4   Conflict Analysis of the Infrastructure Development Conflict

4.1   Modeling the Conflict

Decision Makers and options available to DMs in the Conflict: All parties in CPEC have a consensus on the importance of each railway route. However, the whole conflict revolves around prioritizing the eastern route over a western and central route. Taking into consideration the background of the route controversy and perspective of all stakeholders, we can identify four main players in this conflict. The GOP (DM1), the Government of Baluchistan (DM2), the Government of KPK (DM3) and the Government of China (DM4). It is pertinent to mention that Punjab and Sindh have the same opinion as GOP. The rationale for such consideration lies in the fact that Eastern Route covers most areas in Punjab and Sindh provinces.

Due to its strategic and economic values and requirements of less radical changes in the existing infrastructure, both provinces support the Federal Government. So, in this case these two provinces and the GOP will be considered as one decision-maker (DM1). The planning of railways and land acquisition is largely related to federal and provincial governments, thus the Power Interest Matrix of legitimate stakeholders in the current scenario, shows that the influence of NGO's and Industry is minimal. Civil Society's political power and interest were well presented in All Parties Conferences held in 2015 and 2016. As the Government of KPK and Government of Baluchistan were upholding the same opinion as resolved in APC's, it can be inferred that they are representing civil society.

Under given scenario, the DM1 has four options – Eastern route (ER), Central Route-I (CR-I), Central Route-II (CR-II), and Western Route (WR) (see Table 1).

The DM2 demands connecting Gwadar to the existing infrastructure and upgrading railway infrastructure on the priority basis (Table 1). They are in favor of the construction of Western alignment first (the option DWR in Table 1. The DM3 demands revision of routes in such a way that in case of war, Pashtun belt areas would benefit from it. Therefore, they demand the construction of western alignment on the priority basis (Revise the Project Priorities (Rev) in Table 1)). The DM4 has 2 options. China wants to resolve the conflict and wants completion of EHPs as soon as possible so they will favor the construction of option 1 (Resolve & Implement (R&I) in Table 1). This will be referred to as Option 7. On the other hand, China may also accept any alternative route (AR) as agreed by Pakistani Government.

Table 1. Options of the Decision-makers

DM	Options
Federal (DM1)	1. *Eastern route (ER):* Upgrade the ML-1 and link to China (Fig. 1)
	2. *Central route-I (CR-I):* Upgrade ML-2 and link to China
	3. *Central route-II (CR-II):* Gwadar-Turbat-Panjgur-Basima-Jacobabad section and Jacobabad-Attock Section of ML-2 and link to China (Fig. 1)
	4. *Western route (WR):* The WR as four parts;
	a. Construct Gwadar-Turbat-Panjgur-Basima-Kalat-K-Spezand
	b. Upgrade Spezand-Quetta-Bostan-Muslim Bagh-Qila Saifullah-Zhob section
	c. Construct Zhob-Dera Ismail Khan-Darya Khan section
	d. Upgrade DI Khan-Darya Khan-Kundian-Attock section of ML-2 and link to China (Fig. 1)
Baluchistan (DM2)	5. *Demand WR (DWR):* Baluchistan demands connection of Gwadar to existing and upgraded railway infrastructure on the priority basis
KPK (DM3)	6. *Revise the Project Priorities (Rev):* KPK wants the project plan to be revised. Priority should be given to the western route in Implementation of the project
China (DM4)	7. *Resolve & Implement (R&I):* Favor eastern route under MoU of EHP
	8. *Alternative Routes (AR):* Consider and favor alternative railway routes

Feasible States: Having 4 DMs and 8 options in total, there would be 256 states mathematically. But all these states are not reasonably feasible. For instance, the state NNNYNNNN is not feasible as it cannot be constructed if China is not willing to consider it as an alternative route and prioritize EHP under the CPEC initiative. After deleting the infeasible states, the authors left the 54 reasonable states in the conflict as summarized in Table 2.

Preferences of the DMs: The option statements and state preference of the DMs are given in Table 3. With respect to the state preferences, DM1 would like the construction of the eastern route and the central route-II to be developed and upgraded, with support of China, as EHPs. DM1 covets DM2 and DM3 not to claim the western route and the revision of the project and its priorities, respectively, and to carry out the signed agreement between China and Pakistan. So, it makes the state S47 the most preferred state for the DM1.

Table 2. Feasible States

States	Federal				Bal.	KPK	China	
	1	2	3	4	5	6	7	8
	ER	CR-I	CR-II	WR	DWR	Rev	R & I	AR
S_1	N	N	N	Y	N	N	Y	N
S_2	N	N	N	Y	N	N	Y	Y
S_3	N	N	N	Y	N	Y	Y	N
S_4	N	N	N	Y	N	Y	Y	Y
S_5	N	N	N	Y	Y	N	Y	N
S_6	N	N	N	Y	Y	N	Y	Y
S_7	N	N	N	Y	Y	Y	Y	N
S_8	N	N	N	Y	Y	Y	Y	Y
S_9	N	N	Y	N	N	N	Y	N
S_{10}	N	N	Y	N	N	N	Y	Y
S_{11}	N	N	Y	N	N	Y	Y	N
S_{12}	N	N	Y	N	N	Y	Y	Y
S_{13}	N	N	Y	N	Y	N	Y	N
S_{14}	N	N	Y	N	Y	N	Y	Y
S_{15}	N	N	Y	N	Y	Y	Y	N
S_{16}	N	N	Y	N	Y	Y	Y	Y
S_{17}	N	Y	N	N	N	N	Y	N
S_{18}	N	Y	N	N	N	N	Y	Y
S_{19}	N	Y	N	N	N	Y	Y	N
S_{20}	N	Y	N	N	N	Y	Y	Y
S_{21}	N	Y	N	N	Y	N	Y	N
S_{22}	N	Y	N	N	Y	N	Y	Y
S_{23}	N	Y	N	N	Y	Y	Y	N
S_{24}	N	Y	N	N	Y	Y	Y	Y
S_{25}	N	Y	N	Y	N	N	Y	N
S_{26}	N	Y	N	Y	N	N	Y	Y
S_{27}	N	Y	N	Y	N	Y	Y	N
S_{28}	N	Y	N	Y	N	Y	Y	Y
S_{29}	N	Y	N	Y	Y	N	Y	N
S_{30}	N	Y	N	Y	Y	N	Y	Y
S_{31}	N	Y	N	Y	Y	Y	Y	N
S_{32}	N	Y	N	Y	Y	Y	Y	Y
S_{33}	Y	N	N	N	N	N	Y	N
S_{34}	Y	N	N	N	N	Y	Y	N
S_{35}	Y	N	N	N	N	Y	Y	Y
S_{36}	Y	N	N	N	Y	N	Y	N
S_{37}	Y	N	N	N	Y	Y	Y	N
S_{38}	Y	N	N	N	Y	Y	Y	Y

(*continued*)

Table 2. (*continued*)

States	Federal				Bal.	KPK	China	
	1	2	3	4	5	6	7	8
	ER	CR-I	CR-II	WR	DWR	Rev	R & I	AR
S_{39}	Y	N	N	Y	N	N	Y	N
S_{40}	Y	N	N	Y	N	N	Y	Y
S_{41}	Y	N	N	Y	N	Y	Y	N
S_{42}	Y	N	N	Y	N	Y	Y	Y
S_{43}	Y	N	N	Y	Y	N	Y	N
S_{44}	Y	N	N	Y	Y	N	Y	Y
S_{45}	Y	N	N	Y	Y	Y	Y	N
S_{46}	Y	N	N	Y	Y	Y	Y	Y
S_{47}	Y	N	Y	N	N	N	Y	N
S_{48}	Y	N	Y	N	N	N	Y	Y
S_{49}	Y	N	Y	N	N	Y	Y	N
S_{50}	Y	N	Y	N	N	Y	Y	Y
S_{51}	Y	N	Y	N	Y	N	Y	N
S_{52}	Y	N	Y	N	Y	N	Y	Y
S_{53}	Y	N	Y	N	Y	Y	Y	N
S_{54}	Y	N	Y	N	Y	Y	Y	Y

Based on the option statements the most preferred strategy for DM2 is the state S30 (Table 3) that incorporates the development of CR-I and WR without revising the priorities of the project as claimed by DM3 as it delays the project. DM2 wants DM4 to consider an alternative route option that includes connectivity of Gwadar with main railway infrastructure and to consider it as EHP. Similarly, the same strategy S30 is also the most preferred strategy for DM3 KPK as it would not only provide access to remote areas of KPK but would also link Gwadar with main railway system in EHPs.

The state S31 is the preferred strategy for DM4 China, according to the option statements regarding the construction of CR-I as an EHP, it also involves the rejection of the DM2's aspiration for WR and the DM3's demand for the plan revision.

4.2   Stability Analysis

Having the set of the decision makers, N = {Federal, Baluchistan, KPK, China}, set of all reasonably feasible states X = {S1, S2,...,S54}, reachable lists, and preference rankings of the DMs, the stability analysis has been performed. After the stability analysis of individual states of each DM, state S53 (YNYNYYYN) (see Table 4) is the equilibrium of the conflict because it satisfies all the stability definitions 3–6 for every DM involved in the railway route selection conflict. Therefore, a final solution to a conflict may be an equilibrium if it satisfies all the solution concepts [35]. The equilibrium state S53 (YNYNYYYN) justifies the construction and upgrade of ML-1 (Eastern route) as EHP. It also incorporates construction and upgrade of ML-2 (Central

Table 3. Option Statements and State Preferences of the DMs

DM 1: Federal	
Option statements	Preferences
1 7IFF1 3 2 -4 -5 -6 -8	$S_{47} \succ S_{46} \succ S_{49} \succ S_{48} \succ S_{52} \succ S_{51} \succ S_{54} \succ S_{53} \succ S_{31} \succ S_{34} \succ S_{33} \succ S_{35} \succ$ $S_{37} \succ S_{36} \succ S_{39} \succ S_{38} \succ S_{41} \succ S_{40} \succ S_{43} \succ S_{42} \succ S_{45} \succ S_{44} \succ S_{50} \succ S_{32} \succ$ $S_{10} \succ S_9 \succ S_{12} \succ S_{11} \succ S_{14} \succ S_{13} \succ S_{16} \succ S_{15} \succ S_{18} \succ S_{17} \succ S_{20} \succ S_{19} \succ$ $S_{22} \succ S_{21} \succ S_{24} \succ S_{23} \succ S_{26} \succ S_{25} \succ S_{28} \succ S_{27} \succ S_{30} \succ S_{29} \succ S_2 \succ S_1 \succ S_4 \succ$ $S_3 \succ S_6 \succ S_5 \succ S_8 \succ S_7$

DM 2: Baluchistan	
4 5 8IFF5 3 2 6 7 1	$S_{30} \succ S_{28} \succ S_{45} \succ S_8 \succ S_{43} \succ S_6 \succ S_{26} \succ S_{41} \succ S_4 \succ S_{39} \succ S_2 \succ S_{29} \succ S_{27} \succ$ $S_{44} \succ S_7 \succ S_{42} \succ S_5 \succ S_{25} \succ S_{40} \succ S_3 \succ S_{38} \succ S_1 \succ S_{53} \succ S_{15} \succ S_{51} \succ S_{13} \succ$ $S_{50} \succ S_{23} \succ S_{21} \succ S_{36} \succ S_{35} \succ S_{48} \succ S_{11} \succ S_{46} \succ S_9 \succ S_{19} \succ S_{17} \succ S_{33} \succ S_{32} \succ$ $S_{31} \succ S_{54} \succ S_{16} \succ S_{52} \succ S_{14} \succ S_{24} \succ S_{22} \succ S_{37} \succ S_{49} \succ S_{12} \succ S_{47} \succ S_{10} \succ$ $S_{20} \succ S_{18} \succ S_{34}$

DM 3: KPK	
4 7&8IFF4 5 6 2 3 1	$S_{30} \succ S_{45} \succ S_8 \succ S_{28} \succ S_{43} \succ S_6 \succ S_{26} \succ S_{41} \succ S_4 \succ S_{39} \succ S_2 \succ S_{29} \succ S_{44} \succ$ $S_7 \succ S_{27} \succ S_{42} \succ S_5 \succ S_{25} \succ S_{40} \succ S_3 \succ S_{38} \succ S_1 \succ S_{23} \succ S_{53} \succ S_{15} \succ S_{36} \succ$ $S_{21} \succ S_{50} \succ S_{51} \succ S_{13} \succ S_{35} \succ S_{19} \succ S_{48} \succ S_{11} \succ S_{32} \succ S_{33} \succ S_{17} \succ S_{46} \succ S_9 \succ$ $S_{31} \succ S_{24} \succ S_{54} \succ S_{16} \succ S_{37} \succ S_{22} \succ S_{52} \succ S_{14} \succ S_{20} \succ S_{49} \succ S_{12} \succ S_{34} \succ$ $S_{18} \succ S_{47} \succ S_{10}$

DM 4: China	
1 7IFF1 -6 -8 -5 -4 3IFF8 2IFF8	$S_{31} \succ S_{46} \succ S_{38} \succ S_{35} \succ S_{51} \succ S_{42} \succ S_{47} \succ S_{39} \succ S_{52} \succ S_{43} \succ S_{33} \succ S_{48} \succ$ $S_{40} \succ S_{36} \succ S_{53} \succ S_{44} \succ S_{49} \succ S_{34} \succ S_{41} \succ S_{54} \succ S_{37} \succ S_{45} \succ S_{50} \succ S_{32} \succ$ $S_{17} \succ S_9 \succ S_1 \succ S_{21} \succ S_{13} \succ S_5 \succ S_{27} \succ S_{10} \succ S_{18} \succ S_2 \succ S_{14} \succ S_{22} \succ S_{28} \succ$ $S_6 \succ S_{19} \succ S_{11} \succ S_3 \succ S_{25} \succ S_{23} \succ S_{15} \succ S_7 \succ S_{29} \succ S_{12} \succ S_{20} \succ S_{26} \succ S_4 \succ$ $S_{16} \succ S_{24} \succ S_{30} \succ S_8$

route-II), with its connectivity with Gwadar, satisfying the DM2 & DM3. Moreover, it also fulfills the DM4 aspirations of the completion of EHPs in stipulated time without revising the plan. The equilibrium state also satisfies Chinese stance considering other stakeholders.

The concluded equilibrium strategy is viable and hence helpful in completing the Eastern route (ER) as EHP as decided in the MoU inked between the Chinese and Pakistan governments. The Chinese and Pakistan governments want to complete the

Table 4. Stability analysis and equilibrium

DM	Options	Equilibrium states
Federal (DM1)	1. *Eastern route (ER)*	Y
	2. *Central route-I (CR-I)*	N
	3. *Central route-II (CR-II)*	Y
	4. *Western route (WR)*	N
Baluchistan (DM2)	5. *Demand WR (DWR)*	Y
KPK (DM3)	6. *Revise the Project Priorities (Rev)*	Y
China (DM4)	7. *Resolve & Implement (R&I)*	Y
	8. *Alternative Routes (AR)*	N

ER on time. However, the Central route-II (CR-II) can be considered along with ER in EHP of CPEC project. It would reflect the aspirations of the Baluchistan province to connect Gwadar with the main economic corridor as well as KPK's desire to revise priority route. The Chinese government has not considered the alternative railway routes. The government of Pakistan could negotiate with the Chinese government the development of the CR-IIs. The results were validated using different options and the same state appears to be only feasible state given aspirations decided in APC's and agreed MOU's of CPEC.

5   Conclusion

Collective wisdom reveals the significance of regional economic corridors in enhancing regional development through regional economic cooperation. It is believed to bring prosperity to the participating countries. Initiatives like BRI and CPEC provide vision to act as game changers for the underdeveloped economic regions. The economic corridors not only link the regions with modern infrastructure networks but also provide entrepreneurial and employment opportunities. But, it requires identification of legitimate stakeholders and aggregation of their interests not only at national but also at regional and local levels. The CPEC being a long-term initiative has multiple projects ranging from short-term to medium, and to long-term. Therefore, it needs incorporation of political and social interests at every level of planning and its execution.

In such processes, a dialogue between the politician and the development planners is very important. In this case, it has been observed that the identification of legitimate stakeholders is imminent due to asymmetry in illustration of perspectives, domain of decision-making, concentration of power and resources and information sharing. There was a consensus among stakeholders regarding the initiative but the prioritization of their interests had caused conflict that required collaborative planning at the initial stages.

It has been learned that if such initiatives are treated as a national secret to stakeholders and a façade of inclusion is kept in the planning process, it gives the impression neglect and causes undesirable consequences. Such development projects, in developing countries like Pakistan, are politically motivated by both the ruling party and the opposition, and can jeopardize the implementation of the agreed terms of reference spelled out in international treaties and project execution hampering further FDI. However, political support could ensure smooth project execution without delays.

The projects of that magnitude need creation of strong association of all key stakeholders with unified context in spatial planning. Lack of trust of the regions in the Federal Government could be avoided by building an understanding and including the legitimate stakeholders at planning stage. Intergovernmental negotiation and contracts need to consider all sub-national and regional perspectives. Moreover, sharing of vision and its proper interpretation along with timely dissemination of up to date information could eliminate unwanted conflict based in mistrust.

In the case of CPEC, planners at federal levels also portrayed an autarchic bias having no consultation with provinces. Such planners needed to consider a compassionate perspective regarding both time and resources. Such dominating top to bottom approach of GOP in pursuit of political gain also created an interesting insight into conflict analysis to learn it with attitude perspectives taking into consideration provincial autonomy in signing development plans at their own level in future. The paper enriches the current understanding of negotiation strategy in regional development projects by identifying and applying concepts from GMCR within cross-border regional collaboration. This reflects a growing need to develop system approaches in regional economic development policymaking to provide a win-win solution acceptable to all.

Future research could extend the presented methodology to examining other kinds of collaborative planning efforts in regional economies in public policies, urban planning, setting up economic zones clusters, dealing with environmental and heritage issues, organizational intersegment and planning. Moreover, this study is factual in the sense that it considered real facts but not behaviors. Future researchers could incorporate attitudes of decision makers in such negotiations as strategy. This research also provides grounds for using hyper games for strategic negotiation.

References

1. Li, J., Ma, C., Xie, X.: Research on employment opportunities under the framework of China Pakistan economic corridor, G.C. University, Lahore (2015)
2. ECDPM: European Centre for Development Policy Management, February 2014. http://ecdpm.org/publications/drives-regional-economic-integration-lessons-maputo-development-corridor/. Accessed 11 April 2017
3. Kuklinski, A.: Regional development, regional policies and regional planning. Reg. Stud. **4**(3), 269–278 (1970)
4. Kovacs, E., Fabok, V., Kaloczkai, A., Hansen, H.P.: Towards understanding and resolving the conflict related to the Easter Imperial Eagle (Aquila Heliaca) conservation with participatory management planning. Land Use Policy **54**, 158–168 (2016)
5. Ploger, J.: Strife: urban planning and egonism. Plan. Theory **3**(1), 71–92 (2004)
6. Davies, S.R., Selin, C., Gano, G., Pereira, A.G.: Citizen management and urban change: three case studies of material deliberations. Cities **29**, 351–357 (2012)
7. Fraser, N.M., Hipel, K.W.: Conflict Analysis: Models, and Resolutions. North-Holland, New York (1984)

8. Aas, C., Fletcher, J., Ladkin, A.: Stakeholder collaboration and heritage management. Ann. Tourism Res. **32**(1), 28–48 (2005)
9. Kou-Ann, C., Xi, Z., Fei, Y.: Harmonious urban development and strategic transportation planning in China. In: Seventh International Conference of Chinese Transportation Professional Congress (ICCP), 21–22 May. American Society of Civil Engineers (ASCE), Shanghai (2007)
10. Ruuska, I., Lehtonen, P., Aaltonen, K.: A stakeholder network perspective on unexpected events and their management in international projects. Int. J. Managing Projects Bus. **3**(4), 564–588 (2010)
11. Ali, W., Gang, L., Raza, M.: China-Pakistan economic corridor: current developments and future prospect for regional integration. Int. J. Res. **3**(19), 210–222 (2016)
12. Fang, L., Hipel, K.W., Kilgour, D.M.: Interactive Decision Making: The Graph Model for Conflict Resolution. Wiley, New York (1993)
13. Kilgour, D.M., Hipel, K.W.: The Graph model for conflict resolution: past, present, and future. Group Decis. Negot. **14**, 441–460 (2005)
14. Bilal, S.H.: China-Pakistan economic corridor: regional connectivity. China Int. Bus. **6**, 44–51 (2015)
15. Yunjiao, X., Guowei, Z.: Strengthening energy cooperation between China and Pakistan through CPEC. In: Proceedings of International Conference on CPEC Held at GC University, Lahore, 09–10 December 2015
16. CPEC, China Pakistan Economic Corridor (2017). http://cpec.gov.pk/faqs#. Accessed 04 April 2017
17. BOI: List of Agreements/MoUs Signed during visit of Chinese President, 06 June 2015. http://boi.gov.pk/userfiles1/file/List%20of%20MoUs.docx. Accessed 04 April 2017
18. Mahsud, S.A.: The Pushtun Times.com, January 2016. http://thepashtuntimes.com/wp-content/uploads/2016/01/Working-Paper-of-PUT-on-CPEC.pdf. Accessed 7 April 2017
19. GOB: Government of Baluchistan, May 2015. http://cmpru.gob.pk/reports/CPEC.pdf. Accessed 04 April 2017
20. The Express Tribune: The Express Tribune, 5 December 2016. https://tribune.com.pk/story/1253694/cpecs-western-route-protests-may-erupt-unkept-promises/. Accessed 04 April 2017
21. MOR: Government of Pakistan, 30 June 2016. http://202.83.164.29/railwaysweb/userfiles1/file/ybook201415.pdf. Accessed 04 April 2017
22. Rana, S.: The Express Tribune, 26 July 2015. https://tribune.com.pk/story/926582/economic-corridor-eastern-cpec-route-unfeasible/. Accessed 05 April 2017
23. Ashraf, M.M.: The News International, 9 October 2016. https://www.thenews.com.pk/print/155971-The-CPEC-controversy. Accessed 04 April 2017
24. Haider, I.: Dawn.com, 28 May 2015. https://www.dawn.com/news/1184733. Accessed 04 April 2017
25. Daily Times: Daily Times, 17 January 2016. http://www.dailytimes.com.pk/editorial/17-Jan-2016/cpec-discontents. Accessed 4 April 2017
26. Malik, M.: The Nation, 23 July 2015. http://nation.com.pk/E-Paper/lahore/2015-07-23/page-5. Accessed 6 April 2017
27. The Express Tribune: The Express Tribune, 28 August 2015. https://tribune.com.pk/story/1262963/pm-inaugurates-key-part-cpec-western-corridor-turbat/. Accessed 4 April 2017
28. Abbasi, A.: The News International, 7 October 2016. https://www.thenews.com.pk/print/155573-Propaganda-against-CPEC-upsets-China. Accessed 4 April 2017
29. Pakistan Today: Pakistan Today, 22 November 2015. http://www.pakistantoday.com.pk/2015/11/22/cpec-for-all-two-routes-being-developed-instead-of-one/. Accessed 5 April 2017

30. Ali, S., Xu, H., Xu, P., Zhao, S.: The analysis of environmental conflict in changzhou foreign language school using a hybrid game. Open Cybern. Systemics J. **11**(1), 94–106 (2017)
31. Li, K., Kilgour, D.M., Hipel, K.W.: Status quo analysis of the graph model for conflict resolution. J. Oper. Res. Soc. **56**(6), 699–707 (2005)
32. Nash, J.F.: Equilibrium points in n-person games. Proc. Natl. Acad. Sci. U.S.A. **36**, 48–49 (1950)
33. Nash, J.F.: Non-cooperative games. Ann. Math. **54**, 286–295 (1951)
34. Howard, N.: Paradoxes of Rationality: Theory of Metagames and Political Behavior. MIT, Cambridge (1971)
35. Kassab, M., Hipel, K.W., Hegazy, T.: Conflict resolution in construction disputes using the graph model. Journal of Construction Engineering and Management **132**(10), 1043–1052 (2006)

Attitudinal Analysis of Russia-Turkey Conflict with Chinese Role as a Third-Party Intervention

Sharafat Ali, Haiyan Xu$^{(\boxtimes)}$, Peng Xu, and Michelle Theodora

College of Economics and Management,
Nanjing University of Aeronautics and Astronautics, Nanjing 211106, Jiangsu,
People's Republic of China
xuhaiyan@nuaa.edu.cn

Abstract. The presented attitude-based conflict analysis models the Russia-Turkey conflict with the third-party intervention of China. Third-party intervention model considers the attitudes of three decision makers (DMs) to understand the behaviors of the DMs in decision making in the situation of a strategic conflict. Three sets of attitudes of DMs are considered for attitudinal conflict analysis. The study traces out how the inappropriate (negative) attitudes of Russia and Turkey, regardless of third-party's attitude, would lead to unfavorable consequences. Even though the third-party, China, changes her attitude from neutral to positive, it would not affect the outcome. The attitudinal analysis reveals that the attitude of the focal decision maker, Russia, is important as the change in it influences the outcome of the conflict. The appropriate (positive) attitude of DMs would help resolve the conflict.

Keywords: Strategic conflicts · Attitudinal analysis · Third-party intervention
Russia · Turkey · China

1 Introduction

The interdependence of economies in the 21st century is unprecedented in the human history. The basis of this interdependence is rooted in the economic globalization and expansion in international trade. International trade has made possible the efficient utilization of the global resources and increased levels of well-being and higher levels of consumption. However, despite the interdependence of the countries conflicts are also inevitable. Any unprecedented event of strategic importance happening in one country can influence its relationship with other countries directly or indirectly at different magnitudes. In recent history, Russia-Turkey relations were affected by the Syrian crises. Even though Russia and Turkey shared good economic and diplomatic relations, they have conflicting national and strategic interests in Syria. Russian government supports current regime whereas Turkish government backs the rebels who try to oust Bashar al-Assad [1].

The situation worsened when Turkey shot down a Russian jet near Turkish-Syrian border [2–5]. Despite having good economic relations, this provoked tensions not only

© Springer International Publishing AG, part of Springer Nature 2018
Y. Chen et al. (Eds.): GDN 2018, LNBIP 315, pp. 167–178, 2018.
https://doi.org/10.1007/978-3-319-92874-6_13

between the two countries but also among other adjacent countries which have diplomatic relations with Russia and Turkey. The shooting down of the Russian jet turned into a serious strategic conflict. Both countries showed aggressive behavior to each other. Russian reaction to this provocation could have had serious implications. The Russian-Turkish conflict could have been resolved by considering their attitudes toward each other. Moreover, the consideration of the attitude of any mediating country as a third-party intervention could have also helped resolve the Russia-Turkey conflict.

The resolution of a strategic conflict by using graph model for conflict resolution (GMCR) [6] is insightful as it systematically models a strategic conflict and provides deeper insight with less information as compared to other decision analysis approaches [6–8]. The GMCR is based on classical game theory [9] and the meta-game theory [10]. The behaviors of decision makers (DMs) have important implications on the nature of the conflict [7, 11, 12]. Inohara et al. [12] introduced attitude in the GMCR for the conflict analysis of the war of 1812 between the United States and the UK. However, Inohara et al. generated state prioritizations based on the states' considering attitudes of the DMs. It makes the state prioritization cumbersome when there is a large number of feasible states [7, 13]. Xu et al. [7] introduced attitude-based options to generate the ranking of the states. Matrix representation of general GMCR was introduced in [14–16]. Preference generation based on options makes it convenient to generate states' preferences. Moreover, the attitudinal stability definition presented in Inohara et al. [12] is logical. Walker et al. [17] converted logical attitude methodology into matrix form to improve the ease and efficiency of the attitudinal conflict analysis in the GMCR. Matrix representation provides extended flexibility to attitude calculations and helps encode attitude into a Decision Support System (DSS) [17].

There have been evolutions and improvements in DSS for the GMCR. The first DSS GMCR provided convenience for the use of the GMCR approach and its associated algorithms. DSS GMCR II [18, 19] allowed the users to create their own model to analyze a conflict. It opened the avenues of possibilities for the researchers and decision analysts to analyze complex conflicts in GMCR [17]. The matrix representation of the stability and solution concepts in the GMCR expanded the realm of applicability for the algebraic approach.

The MRCRDSS, based on matrix representation is useful in carrying out the individual stability analysis. The representation of attitudinal stability concepts has been introduced in the MRCRDSS [17]. It makes the incorporation and analysis of multiple decision makers' attitude while analyzing a conflict. The objective of the present study is to analyze the Russia-Turkey conflict while considering the mediating role of China. Moreover, the attitude of the DMs in a three DMs model is incorporated into the MRCRDSS. The study analyzes how the attitude of the intervening third-party may have impacted the outcomes of the conflict. Furthermore, the study also traces out how the attitude and changes in the attitudes of Russia and Turkey with the third-party intervention affected the nature and outcomes of the conflict.

The rest of the paper is structured as follows. Section 2 represents the GMCR and attitudinal stability concepts. Section 3 is comprised of the background of the Russia-Turkey conflict. The results of the stability analysis are summarized and discussed in Sect. 4. The conclusion of the analysis and policy implications is presented in Sect. 5.

2   Attitude-Based Conflict Analysis Under GMCR

2.1   The Graph Model for Conflict Resolution

A GMCR is a 4-tuple; $(K, X, (A_i)_{i \in K}, \succ_i, \sim_i)$. Where K and X, respectively, are the set of all decision makers (DMs) ($|K| \geq 2$) and the set of all states in a conflict. (X, A_i) is the DM i's directed graph with the set of all vertices X and the set of all arcs, $A \subset X \times X$, that are movements controlled by DM i between the pair of states. DM i's preferences on X are denoted by $(\succ_i, \sim_i)$ [6]. The DMs, in a conflict, make moves and counter-moves in order to do what they possibly could do. Therefore, a graph establishes a natural construct to model a conflict in which nodes represent the possible states and the arcs systematically keep track of a given DM's movements that he could make in one step.

For $i \in K$ and $x_1, x_2 \in X$, $x_1 \succ_i x_2$ implies that state x_1 is preferable to x_2 for DM i. Whereas, $x_1 \sim_i x_2$ means that the DM i is indifferent between the two states. The relative preferences are assumed to be asymmetric reflexive, and complete. The preference $\succ_i$ is asymmetric if, for all $x_1, x_2 \in X$, $x_1 \succ_i x_2$ and $x_2 \succ_i x_1$ cannot hold simultaneously. However, $\sim_i$ is symmetric as $x_1 \sim_i x_2$ and $x_2 \sim_i x_1$ can hold simultaneously. Moreover, $(\succ_i, \sim_i)$ is complete as for all $x_1, x_2 \in X$, as one of $x_1, x_2 \in X$, $x_1 \succ_i x_2$, or $x_1 \sim_i x_2$ is true.

2.2   The Attitude of the DMs

The attitude of the DMs towards other DMs, in decision-making, plays a pivotal role in determining their preferences, moves, and counter-moves from one state to another [7, 12, 13]. It is a stable psychological attitude of an individual to particular person, event, idea, or emotion. The attitude contains a subjective evaluation of the individual and the preferences of the DMs, in a conflict, can be generated by subjective evaluation of DMs [7].

Owing to the importance of the attitudinal preferences of the DMs, Inohara et al. [12] considered three kinds of attitude in conflict analysis in their graph model. In recent studies [7, 13], the attitude-based prioritization is used to generate preferences of the states considering positive, negative, and neutral attitudes of the DMs. In option prioritization method, for $i, j \in K$, the DM i's option statement is $O_i(i = 1, 2, \ldots, k)$. Under this option statement, the DM i's *preference*, $P_i(i = 1, 2, \ldots, k)$, can be obtained. Before moving forward to attitudinal stability concepts some definitions [7] need to be summarized as follows:

Definition 1: Option Statement with Positive Attitude: For $i, j \in K$, $O_i(a_{ij} = +) = O_j$. Where a_{ij} indicates the attitude of the DM i towards the DM j. Having the positive attitude, the DM i's option statement would be same as the DM j's option statement.

Definition 2: Option Statement with Negative Attitude: For $i, j \in K$, $O_i(a_{ij} = -) = -O_j$. It implies that if DM i has negative attitude for DM j (*i.e.* $a_{ij} = -$), her option statement would be opposite of the DM j's option statement. That would not be beneficial for the DM j.

Definition 3: Option Statement with Neutral Attitude: For $i,j \in K$, $O_i(a_{ij} = 0) = I$. Where "I" stands for indifferent. It means if DM i has a neutral attitude towards her opponent, she does not care about the option statement of the opponent.

Definition 4: Attitude Preference: According to the option statement, the attitude preference of DM i (T_{ij}) can be obtained. For $i,j \in K$, and $x_1, x_2 \in X$, $x_2 \in T_{ij}(x_1)$ if and only if (IFF) $x_2 \succ_i x_1$ satisfies T_{ij}.

Definition 5: Total Attitude Preference: For $i,j \in K$, and $x_1, x_2 \in X$, $x_2 \in T_i^+(x_1)$ IFF $x_2 \in T_{ij}(x_1)$ for all $j \in K$. Total attitude preference of DM i is the intersection of all her attitude preferences that she wants to reach.

Definition 6: Set of Less or Equally Preferred States at Total Attitude: The set of all less or the equally preferred states at total attitude for DM i, for $i \in K$, is $x_2 \in T_i^{-=}(x_1)$ IFF $x_2 \notin T_i^+(x_1)$.

Definition 7: Reachable List: The reachable list for DM i, for $i,j \in K$ and $x_1, x_2 \in X$, from state x is the set $\{x_2 \in X | (x_1, x_2) \in A_i\}$. It can be symbolized as $R_i(x) \subset X$.

Definition 8: Unilateral Improvement (UI) List: The UI list for DM I, for $i,j \in K$ and $x_1, x_2 \in X$, is $x_2 \in T_i^*(x_1)$ IFF $x_2 \in R_i(x_1)$ and $x_2 \in T_i^+(x_1)$.

2.3   Attitude-Based Stability Definitions

Definition 9: Relational Nash Stability (RNASH): A state x is RNASH stable for DM i IFF there is no UI for her. Symbolically, for $i,j \in K$ and $x_1, x_2 \in X$, $x \in X_i^{RNASH}$ IFF $T_i^*(x) = \phi$. In this case, the DM i has no incentive to move from state x.

Definition 10: Relational General Metarationality (RGMR): A state x_1 is RGMR for DM i, for $i,j \in K$, if for all $y \in T_i^*(x)$ and $R_i(y) \cap T_i^{-=}(x) \neq \phi$, denoted by $x \in X_i^{RGMR}$. In this case, the DM i would not move to UI state at an attitude if keeping in mind her opponent could sanctions her move irrespective of benefit to herself.

Definition 11: Relational Symmetric Metarationality (RSMR): A state x is RSMR, $x \in X_i^{RSMR}$, if for all $y \in T_i^*(x)$, there exists $z \in R_i(y) \cap T_i^{-=}(x)$ and $m \in T_i^{-=}(x)$ for all $m \in R_i(z)$. According to RSMR stability concept, if DM i could not avoid sanction on her UI moves at an attitude by the opponent then she would not move from state x. This makes state x RSMR stable for DM i.

Definition 12: Relational Sequential Stability (RSEQ): A state x is RSMR, $x \in X_i^{RSEQ}$, if for all $y \in T_i^*(x)$, and $T_j^*(y) \cap T_i^{-=}(x) \neq \phi$. In RSMR, the DM i's UI moves at attitude are sanctioned by DM j's UI moves. The RSEQ is similar to RGMR except that the DM i considers her own benefit at the time of sanction by her opponent.

2.4   Decision Support System MRCRDSS

The development of a decision support system (DSS) was necessary for the analysis of conflicts with multiple DMs. The MRCRDSS system was developed based on the matrix representations of the GMCR [14–16] and attitudinal matrix representation [17]. The matrix representation of attitude in GMCR has made possible the encoding and therefore development of MRCRDSS for the attitudinal analysis of multiple decision maker conflict. The MRCRDSS is an efficient tool for the analysis. The attitudinal analysis in the present study with three decision makers has been carried out using this DSS.

3   Background of the Russia-Turkey Conflict

3.1   The Russia-Turkey Conflict

The relationship between Russia and Turkey goes centuries back and is complicated in nature. However, the economic and political relations between the two countries became strong after the end of the Cold War and with the emergence of globalization [20, 21]. Turkey has been ranked as the leading trading partner of Russia. In addition to this, Turkey became one of the best destinations for Russian tourists. Turkish business started to flourish in Russia. The politico-economic relations between the two countries became so pleasant that they instituted visa-free travel. But this economic edifice hampered with the conflicting interests over Syrian issue [21]. Turkey shot down a Russian fighter plane near the Turkish-Syrian border [2, 3, 21]. The Russian government showed an inflammatory reaction. Russia imposed heavy trade sanctions on Turkey coupled with a ban on Russian tourism. Moreover, Turkish business and investment in Russia were also adversely affected. Consequently, the situation became worse.

Russia proclaimed that the fighter jet was not in the Turkish airspace but the Turkish version was corroborated by the NATO. The Turkish government was seeking support from the US and the NATO on the issue [22]. The Russian government could have opted to investigate the matter and wait till the findings of investigations were unveiled. But the Russian government immediately imposed the sanctions on Turkey. The sanctions hit the Turkish trade, tourism and construction sectors, and exchanges that were benefitting Turkey. Turkish exports of vegetables and fruit were banned in Russia [23]. However, the Russian government did not reduce the gas supplies to Turkey that accounts for 55% of the total gas consumption [22] and 35% of oil [23]. Russia held Turkey responsible for the incident and demanded apology and payment of indemnities [22].

3.2   Chinese Government Intervention in the Russia-Turkey Conflict

The conflict had serious implications not only for both countries but also for the other countries in the region, especially China. The latter not only possesses greater strategic and economic power but also has strong economic and diplomatic relations with both countries. China could have played a very pivotal role in mediating the Russian-Turkey conflict. The present study models the Russian-Turkish conflict considering their attitude towards each other, while cogitating the intervening role of China as a third-party.

4 Attitudinal Conflict Analysis of the Russia-Turkey Conflict with Third-Party Intervention by China

4.1 Modeling of the Russia-Turkey Conflict with Third Party Intervention by China

The DMs and Options of the DMs: Due to direct participating nature, Russia and Turkey are deemed to be the major DMs in the conflict. Each decision maker has a set of options. Due to the capacity of the Chinese government as the mediator, a set of options was also considered in the analysis. So, there are three DMs in the conflict. When the DMs interact, these sets of options of the DMs are considered as the state strategies. The options of the DMs are summarized in Table 1. The Russian government has two options; to impose economic sanctions on Turkey or investigate further into the matter and then decide how to react against the opponent. The Turkish government also has two options, one is to apologize to Russia (The Russian government asked Turkey to categorically apologize and pay indemnities [22]). The second option is for Turkey to ask the US and the NATO for their support. However, China as the third-party in the conflict has the option to play its role as a mediator.

Table 1. Options of the DMs and the feasible states

Russia													
1. Sanction	N	N	N	N	N	N	N	N	N	Y	Y	Y	Y
2. Investigation	N	N	N	N	N	Y	Y	Y	Y	N	N	N	N
Turkey													
3. Apologize	N	N	N	N	Y	N	N	N	N	N	N	N	N
4. USA's help	N	N	Y	Y	N	N	N	Y	Y	N	N	Y	Y
China													
5. Mediation	N	Y	N	Y	N	N	Y	N	Y	N	Y	N	Y
Label	x_1	x_2	x_3	x_4	x_5	x_6	x_7	x_8	x_9	x_{10}	x_{11}	x_{12}	x_{13}

With three DMs in the modeled conflict and 5 options in total, mathematically, there are 32 states. But due to the mutually exclusive nature of some options and infeasibility of some states, the authors are left with the 13 feasible states. These states, for the sake of simplicity, are labeled as $x_1, x_2, \ldots, x_{13}$.

Option Statements: The options statements in Table 2 show that the Russian government prefers that Turkey apologize and therefore there will be no sanctions. The option statement, 3, −1, 5, 2, −4, describes the preferences of the Russian government from the most preferred to the least. However, Turkey wants Russia not to impose economic sanctions as it would adversely affect the Turkish economy. Turkish government does not like any further investigation into the matter. Chinese intervention for the resolution of the conflict is also the least preferred option for Turkey. China, the intervening third party, also prefers Turkish apology over economic sanctions.

Table 2. Options statement

Russia	Turkey	China
3	−1	3
−1	−2	−1
5	−3	−2
2	4	−4
−4	5	5

Attitudes of the DMs: The stability analysis of the feasible states has been carried out while considering different attitudes (e) of the DMs – Russia (R), Turkey (T), China (C). Three attitude matrices have been considered for the stability analysis:

$$
Attitude\ Matrix - I =
\begin{bmatrix}
e_{RR} = + & e_{RT} = - & e_{RC} = 0 \\
e_{TR} = - & e_{TT} = + & e_{TC} = 0 \\
e_{CR} = 0 & e_{CT} = 0 & e_{CC} = +
\end{bmatrix}
\tag{1}
$$

$$
Attitude\ Matrix - II =
\begin{bmatrix}
e_{RR} = + & e_{RT} = - & e_{RC} = 0 \\
e_{TR} = - & e_{TT} = + & e_{TC} = 0 \\
e_{CR} = + & e_{CT} = 0 & e_{CC} = +
\end{bmatrix}
\tag{2}
$$

$$
Attitude\ Matrix - III =
\begin{bmatrix}
e_{RR} = + & e_{RT} = 0 & e_{RC} = 0 \\
e_{TR} = - & e_{TT} = + & e_{TC} = 0 \\
e_{CR} = + & e_{CT} = 0 & e_{CC} = +
\end{bmatrix}
\tag{3}
$$

In the first attitude matrix (1), the attitude of the Russian government towards herself is positive, towards Turkey is negative and towards China is neutral. Whereas, the Turkish attitude towards herself is positive, towards Russia is negative, and towards China neutral. The negative attitude of Russia and Turkey towards each other is considerable because it is a matter of national integrity and sovereignty for both countries. From Turkey's point of view, Russian fighter jets intruded the Turkish airspace and violated the territorial integrity despite the warning by the Turkish air force [4, 5]. The Russian side argues that the jets were not in the Turkish territory rather they were in the Syrian territory throughout the mission and they did not violate the Turkish airspace; also that no warning was received from the Turkish side. So, the Russian government showed aggressiveness towards Turkey [2, 3, 5]. However, the Chinese government's attitude towards Russia and Turkey is considered neutral.

In the second attitude matrix (2), the attitudes of Russia and Turkey are considered the same but the attitude of the intervening third-party – China is considered to be changing from neutral to positive. The Chinese positive attitude towards Russia is also realistic because of the strong economic and strategic relations between China and Russia. These changes in Chinese attitudes are assumed to be neutral and/or positive in the assessment of the impact of attitude on the overall outcome of the conflict. The assessment of the impact of changes in the attitude of the third part may provide some insights.

In the third attitude matrix (3), the researchers used their freedom to hypothesize a change in Russian attitude from negative to positive towards Turkey. The positive attitude of the Russian government towards Turkey is considered here to analyze whether it affects the equilibrium outcome and helps to resolve the conflict.

4.2 Attitudinal Stability Analysis with Third-Party Intervention

Stability Analysis with Attitude Matrix-I: In the attitude matrix-I, it is assumed that Russia has a positive attitude for herself but a negative attitude for her opponent Turkey. On the other hand, Turkey has a negative attitude towards Russia and positive for herself. However, China's attitude is considered neutral for both of the opposing DMs Russia and Turkey. In this case, the stability analysis results, shown in Table 3, unfold state x_9 and x_{13} as equilibrium states. These states are relational stable under all the stability definitions considered in the analysis.

Table 3. Stability analysis with attitude matrix-I

States	RNASH				RGMR				RSMR				RSEQ			
	R	T	C	Eq	R	T	C	Eq	R	T	C	Eq	R	T	C	Eq
x_1					✓	✓			✓	✓			✓	✓		
x_2			✓		✓	✓	✓	✓	✓	✓	✓	✓	✓	✓		
x_3		✓			✓	✓			✓	✓			✓	✓		
x_4		✓	✓		✓	✓	✓	✓	✓	✓	✓	✓	✓	✓		
x_5	✓		✓		✓	✓	✓	✓	✓	✓	✓	✓	✓	✓	✓	✓
x_6	✓				✓	✓	✓	✓	✓	✓	✓	✓	✓	✓	✓	✓
x_7	✓		✓		✓	✓	✓	✓	✓	✓	✓	✓	✓			✓
x_8	✓	✓			✓	✓	✓	✓	✓	✓	✓	✓	✓	✓		
x_9	✓	✓	✓	✓	✓	✓	✓	✓	✓	✓	✓	✓	✓	✓	✓	✓
x_{10}	✓				✓	✓	✓	✓	✓	✓	✓	✓	✓	✓	✓	✓
x_{11}	✓		✓		✓	✓	✓	✓	✓	✓	✓	✓	✓			✓
x_{12}	✓	✓			✓	✓			✓	✓			✓	✓		
x_{13}	✓	✓	✓	✓	✓	✓	✓	✓	✓	✓	✓	✓	✓	✓	✓	✓

The state x_9 (NYNYY) implies that Russia does not impose the sanction against Turkey and wait until further investigations into the matter. In the meanwhile, Turkey seeks support from the US and other NATO members. In addition to this, China plays mediation role as a third party to resolve the conflict between Russia and Turkey. The equilibrium state x_{13} (YNNYY) is a rather unfavorable outcome. This equilibrium strategy of the conflict implies that the Russian government imposes the sanctions on the Turkish economy. In the meanwhile, Turkish government seeks support from the US, NATO and China asks the two opponents to resolve the issue.

Stability Analysis with Attitude Matrix-II: In the second scenario, the attitudes of the Russian and Turkish governments for themselves and for the opponent are unchanged but the attitude of China – the third-party is changed from neutral to positive towards Russia. In this situation, the stability analysis results, shown in Table 4, reveal the same states as equilibrium states. This indicates that if the opposing decision makers do not change their attitude then change in the attitude of the intervening third party may not have a significant impact on the outcome of the conflict.

Table 4. Stability analysis with attitude Matrix-II

States	RNASH				RGMR				RSMR				RSEQ			
	R	T	C	Eq	R	T	C	Eq	R	T	C	Eq	R	T	C	Eq
x_1					√	√			√	√			√	√		
x_2			√		√	√	√	√	√	√	√	√	√	√		
x_3		√			√	√			√	√			√	√		
x_4		√	√		√	√	√		√	√	√	√	√	√		
x_5	√		√		√	√	√		√	√	√	√	√	√	√	√
x_6	√				√	√	√		√	√	√	√	√	√	√	√
x_7	√		√		√	√	√		√	√	√	√	√			√
x_8	√	√			√	√	√		√	√	√	√	√	√		
x_9	√	√	√	√	√	√	√	√	√	√	√	√	√	√	√	√
x_{10}	√				√	√	√		√	√	√	√	√	√	√	√
x_{11}	√		√		√	√	√		√	√	√	√	√			√
x_{12}	√	√			√	√			√	√			√	√		
x_{13}	√	√	√	√	√	√	√	√	√	√	√	√	√	√	√	√

Stability Analysis with Attitude Matrix-III: The stability analysis in the above two cases, while changing the attitude of the mediating third party, China, from neutral to positive towards Russia has no significant impact on the outcome of the conflict. In the third case, the authors analyze the stability of the states for each decision maker in the conflict considering the change of Russian attitude towards Turkey from negative to neutral. Moreover, the Turkish government's attitude towards Russia is unchanged (*i.e.* negative) and intervening third party's attitude towards Russia is positive but neutral towards Turkey.

The stability analysis results, with third attitude matrix, reveal state x_8 and x_9 as equilibrium states (see Table 5). The equilibrium state x_8 (NYNYN) describes the strategy in which the Russian government does not impose restrictions and sanctions on Turkish economy but awaits the findings of in-depth investigations. The only difference between the state x_8 and x_9 is that, in state x_9 China plays its intervening role in an effort to resolve the conflict.

Table 5. Stability analysis with attitude Matrix-III

States	RNASH				RGMR				RSMR				RSEQ			
	R	T	C	Eq	R	T	C	Eq	R	T	C	Eq	R	T	C	Eq
x_1					✓	✓			✓	✓			✓	✓		
x_2		✓			✓	✓	✓	✓	✓	✓	✓	✓	✓	✓		
x_3		✓			✓	✓			✓	✓			✓	✓		
x_4		✓	✓		✓	✓	✓	✓	✓	✓	✓	✓	✓	✓		
x_5	✓		✓		✓	✓	✓	✓	✓	✓	✓	✓	✓	✓	✓	✓
x_6	✓				✓	✓	✓	✓	✓	✓	✓	✓	✓		✓	
x_7	✓		✓		✓	✓	✓	✓	✓	✓	✓	✓	✓		✓	
x_8	✓	✓	✓	✓	✓	✓	✓	✓	✓	✓	✓	✓	✓	✓	✓	✓
x_9	✓	✓	✓	✓	✓	✓	✓	✓	✓	✓	✓	✓	✓	✓	✓	✓
x_{10}					✓	✓			✓	✓			✓	✓		
x_{11}		✓			✓	✓			✓	✓			✓	✓		
x_{12}		✓			✓				✓				✓			
x_{13}		✓	✓		✓	✓			✓	✓			✓	✓		

5   Conclusion

The present study models the Russia-Turkey conflict which was triggered by shooting down of Russian fighter jet by Turkish forces near Syrian-Turkish border. This incident put Russia and Turkey on the path of hostility that could escalate into military confrontation. The escalated confrontation between the two countries would not only have serious politico- economic and strategic implications affecting their own relations and economic ties but also the other countries in the region and other trading partners with these two economies. Conflicts need to be resolved to avoid undesirable and unfavorable outcomes. Conflict analysis while considering the behavior of decision makers and players could be helpful in understanding the decision-making behavior of the DMs and thereby in understanding the nature and evolution of the conflict. This conflict analysis is an attempt to analyze the Russia-Turkey conflict, by considering their attitudes towards each other and the third-party intervention by China in the framework of the GMCR. This study uses three attitude matrices to examine how different attitudes of DMs affect the outcome of the conflict.

The results of the attitudinal stability analyses unfold that when Russia and Turkey have a negative attitude towards each other, the equilibrium outcomes are not favorable. Even if the intervening third-party, China, changes her attitudes from neutral to positive for Russia (attitude matrix-II), the equilibrium outcomes are not different form the first attitude matrix. However, when the focal decision maker – Russia changes her attitude from negative to positive for Turkey and the intervening third-party, China, has positive attitude towards Russia, the equilibrium outcomes are more favorable. It implies that the attitudes of the DMs in the Russia-Turkey conflict are critical. The point worth noting is that the attitude of the third-party has no effect on the outcome of the conflict unless the focal DM Russia changes her attitude. In this conflict, the attitude of the focal decision maker plays a pivotal role and a positive attitude of the Russian government towards Turkey could be helpful to diffuse the escalated situation and avoid economic repercussions.

References

1. Stubbs, J., Solovyov, D.: Kremlin says Turkey apologized for shooting down Russian jet. Reuters, 27 June 2016. https://www.reuters.com/article/us-russia-turkey-jet/kremlin-says-turkey-apologized-for-shooting-down-russian-jet-idUSKCN0ZD1PR. Accessed 01 Nov 2018

2. Melvin, D., Marinez, M., Bilginsoy, Z.: Putin calls jet's downing 'stab in the back'; Turkey says warning ignored. CNN, 24 November 2015. http://www.cnn.com/2015/11/24/middleeast/warplane-crashes-near-syria-turkey-border/index.html. Accessed 11 Jan 2018

3. Melvin, D., Mullen, J., Bilginsoy, Z.: Tensions rise as Russia says it's deploying anti-aircraft missiles to Syria. CNN, 25 November 2015. http://www.cnn.com/2015/11/25/middleeast/syria-turkey-russia-warplane-shot-down/index.html. Accessed 11 Jan 2018

4. Naylor, H., Roth, A.: NATO faces new Mideast crisis after downing of Russian jet by Turkey. The Washington Post, 24 November 2015. https://www.washingtonpost.com/world/turkey-downs-russian-military-aircraft-near-syrias-border/2015/11/24/9e8e0c42-9288-11e5-8aa0-5d0946560a97_story.html?utm_term=.7251c00b23d3. Accessed 05 Jan 2018

5. BBC: Turkey's downing of Russian warplane - what we know. BBC News, 01 December 2015. http://www.bbc.com/news/world-middle-east-34912581. Accessed 02 Jan 2018

6. Fang, L., Hipel, K.W., Kilgour, D.M.: Interactive Decision Making: The Graph Model for Conflict Resolution. Wiley, New York (1993)

7. Xu, P., Xu, H., He, S.: Evolutional analysis for the South China sea dispute based on the two-stage attitude of Philippines. In: Schoop, M., Kilgour, D.M. (eds.) GDN 2017. LNBIP, vol. 293, pp. 73–85. Springer, Cham (2017). https://doi.org/10.1007/978-3-319-63546-0_6

8. Ali, S., Haiyan, X., Peng, X., Zhao, S.: The analysis of environmental conflict in Changzhou foreign language school using a hybrid game. Open Cybern. Syst. J. **11**, 94–106 (2017)

9. Neumann, V., Morgenstern, J.: Theory of Games and Economic Behavior. Princeton University Press, Princeton (1944)

10. Howard, N.: Paradoxes of Rationality. MIT Press, Cambridge (1971)

11. Kilgour, D.M., Hipel, K.W., Fang, L.: The graph model for conflicts. Automatica **23**(1), 41–55 (1987)

12. Inohara, T., Hipel, K.W., Walker, S.: Conflict analysis approach for investigating attitude and misperceptions in the war of 1812. J. Syst. Sci. Syst. Eng. **16**(2), 181–201 (2007)

13. Xu, H., Xu, P., Ali, S.: Attitude analysis in process conflict for C919 aircraft manufacturing. Trans. Nanjing Univ. Aeronaut. Astronaut. **34**(2), 1–10 (2017)

14. Xu, H., Kilgour, D.M., Hipel, K.W.: Matrix representation of solution concepts in graph models for multiple decision makers graphs. IEEE Trans. Syst. Man Cybern. A Syst. Humans **39**(1), 96–108 (2009)

15. Xu, H., Kilgour, D.M., Hipel, K.W., Kemkes, G.: Using matrices to link conflict evolution and resolution in a graph model. Eur. J. Oper. Res. **207**, 318–329 (2010)

16. Xu, H., Li, K.W., Hipel, K.W., Kilgour, D.M.: A matrix approach to status quo analysis in the graph model for conflict resolution. Appl. Math. Comput. **212**(2), 470–480 (2009)

17. Walker, S.B., Hipel, K.W., Xu, H.: A matrix representation of attitudes in conflicts. IEEE Trans. Syst. Man Cybern. Syst. **43**(6), 1328–1342 (2013)

18. Fang, L., Hipel, K.W., Kilgour, D.M., Peng, X.: A decision support system for interactive decision making - part I: model formulation. IEEE Trans. Syst. Man Cybern. C Appl. Rev. **33**(1), 42–55 (2003)

19. Fang, L., Hipel, K.W., Kilgour, D.M., Peng, X.: A decision support system for interactive decision making, part 2: model formulation. IEEE Trans. Syst. Man Cybern. C **33**(1), 56–66 (2003)

20. Kirisci, K.: Turkey and its post-Soviet neighborhood, vol. 112(756), p. 271 (2013)

21. Kirişci, K.: Order from chaos: The implications of a Turkish-Russian rapprochment, 08 October 2016. https://www.brookings.edu/blog/order-from-chaos/2016/08/10/the-implications-of-a-turkish-russian-rapprochement/. Accessed 05 Jan 2018

22. Özel, S.: The crisis in Turkish-Russian Relations. Center for American Progress 10 May 2016. https://www.americanprogress.org/issues/security/reports/2016/05/10/137131/the-crisis-in-turkish-russian-relations/. Accessed 05 Jan 2018

23. Skinner, A.: Grudge between Ankara and Moscow deepens in struggle for regional influence. CNBC 14 Mar 2016. https://www.cnbc.com/2016/03/14/turkey-v-russia-grudge-between-ankara-and-moscow-deepens-in-struggle-for-regional-influence.html. Accessed 02 Jan 2018

Behavioral Modeling of Attackers Based on Prospect Theory and Corresponding Defenders Strategy

Ziyi Chen, Chunqi Wan, Bingfeng Ge, Yajie Dou[(⊠)], and Yuejin Tan

College of Systems Engineering, National University of Defense Technology,
Changsha, Hunan 410073, China
chenziyi13@nudt.edu.cn, yajiedou_nudt@163.com

Abstract. Many methods focused on describing the attackers' behavior while ignoring defenders' actions. Classical game-theoretic models assume that attackers maximize their utility, but experimental studies show that often this is not the case. In addition to expected utility maximization, decision-makers also consider loss of aversion or likelihood insensitivity. Improved game-theoretic models can consider the attackers' adaptation to defenders' decisions, but few useful advice or enlightenments have been given to defenders. In this article, in order to analyze from defenders' perspective, current decision-making methods are augmented with prospect theory results so that the attackers' decisions can be described under different values of loss aversion and likelihood insensitivity. The effects of the modified method and the consideration of upgrading the defense system are studied via simulation. Based on the simulation results, we arrive at a conclusion that the defenders' optimal decision is sensitive to the attackers' levels of loss aversion and likelihood insensitivity.

Keywords: Prospect theory · Decision analysis · Game theory

1 Introduction

Classic game-theoretic models often assume that attackers and defenders are rational in the sense of the von Neumann-Morgenstern expected utility theory [1]. Methods proposed by Parnell et al. [2] and Rios Insua et al. [3] combine decision analysis and game theory to allow some deviations from the rationality. However, Simon [4] and Kahneman and Tversky [5], have already show that decision-makers are not strictly rational when they engage in unaided decisions. Similarly, players in games deviate from the rationality principles in such ways such as being averse to losses, having diminishing sensitivity, being dependent on reference points, and distorting probability when they face uncertain outcomes. Therefore, descriptive models have been proposed predict decision-makers' actual behavior. Merrick and Parnell [6] give a review of methods used in attacker/defender models to represent the adaptation of the attacker to the defender's decisions. In addition to traditional decision analysis, game-theoretic models and hybrid methods, prospect theory has been widely applied [7, 8]; for example, Mazicioglu and Merrick [9] represented adversaries with multiple objectives

© Springer International Publishing AG, part of Springer Nature 2018
Y. Chen et al. (Eds.): GDN 2018, LNBIP 315, pp. 179–189, 2018.
https://doi.org/10.1007/978-3-319-92874-6_14

in counter-terrorism using multi-attribute prospect theory. Edouard [10] modeled terrorists' behavior with modified decision weights and proposed the strategic logit risk analysis (SLRA) method to solve allocation of scarce defense resources. Merrick [11] modified existing methods used in counter-terrorism decisions, in particular, the attacker's decision problem, with a descriptive model that accounts for the attacker's loss aversion.

An effective confrontation decision framework incorporates the defender's decisions and the attacker's adaptation to them. The literatures mentioned above consider using improved model to describe the behavior of the attacker more realistically, but the people's research is to provide advice to the defender, so we should consider how the defender can arrive the optimal result by changing his behavior. Ideally, this article introduces a "Wait and see" region through a random upgrade approach, which means that defender can actively change their behavior to achieve optimal results based on the type of the attacker's behavior.

The organization of this paper is organized as follows. In the next section, a brief introduction of prospect theory is presented. A comprehensive description of the model is given in Sect. 3. In order to provide advice to the defender the impact of changes in the behavior of the defender, confidential information, and random promotion is considered in Sect. 4. Finally, conclusions and suggestions are given in Sect. 5.

2   Brief Introduction of Prospect Theory

It is a well-known fact that human beings are not purely rational (and not entirely irrational or arbitrary) without the use of decision analysis methods. Therefore, we extend the method of Parnell et al. [2] and Rios Insua et al. [3] to the descriptive model of attackers' decision making so that attackers follow rationality in decision making. First, we will use prospect theory to model attackers' decisions behavior. Prospect theory has been shown to represent loss aversion, framing effects, deterministic effects, and likelihood insensitivity. A prospect is a gamble that has a probability p to get x and a probability $1 - p$ to get 0. According to original prospect theory [4], we set this prospect as $\pi(p)v(x)$, where v is a S-shaped value function, with its inflection point(the reference point) at 0. Thus, it represents diminishing sensitivity and reference dependence. It is also defined that $v(x) - v(0) < v(0) - v(-x)$ represents loss aversion. The probability weighting function π also includes reference dependence for $p = 0$ and $p = 1$, diminishing sensitivity away from these references. Function $v(x)$ is concave for low probability values and convex for high values.

For the more complex situation with more than two outcomes, the original prospect theory cannot obey stochastic dominance. Tversky and Kahneman [8] introduced the now standard form of prospect theory to overcome this shortcoming. They placed the weights on the cumulative probabilities, or ranked probabilities. To apply this form of prospect theory to a prospect X with n outcomes, we first order the outcomes $x_1 \geq \ldots \geq x_k \geq 0 \geq x_{k+1} \geq \ldots \geq x_n$ along with their respective probabilities $p_1, \ldots, p^n$. Note that there are k outcomes which are gains and $n - k$ outcomes which are losses. If $k = 0$, then all outcomes are losses; if $k = n$, then all outcomes are gains. In the improved form of prospect theory, weights are applied to cumulative probabilities:

$$w^+(x_j) = \pi(P(X \geq x_j)) - \pi(P(X > x_j)) \tag{1}$$

for gains $(j \leq k)$, and

$$w^-(x_j) = \pi(P(X \leq x_j)) - \pi(P(X < x_j)) \tag{2}$$

for losses $(j \leq k)$. Thus, the value of a prospect is:

$$\sum_{j=1}^{k} w^+(x_j)v(x_j) + \sum_{j=k+1}^{n} w^-(x_j)v(x_j). \tag{3}$$

In the following example, we will use this prospect theory.

3 Behavioral Modeling of Attackers

A-type missile is a new type of weapon with a nuclear warhead, which can effectively break through the defenses of the current conventional defense systems. C-type defense system is designed specifically for A-type missile, which can effectively intercept its attacks. We use d1 represent the defender decided to upgrade to the C-type defense system, and d2 for not upgrade. The attacker decision represents the choice of attacks using the A-type missile, a1, and attacks using a conventional weapon (not missile), a2. The loss caused by the A-type missile attack is assumed to be 40 billion dollars; parameter r is used to represent the loss. The loss caused by a conventional weapon attack is assumed to be a quarter of this amount.

The cost of upgrading the defense system is assumed to be $100 million. We assume that the attackers want to inflict the largest damage on the defender, but they also suffer losses if the defenders' defense systems intercept their attacks. For illustrative purposes, this loss, f, is assumed to be one-tenth of the value of the successful A-type missile attack. The probability p of successful interception of a Type A missile by the C-type defense system is assumed at 0.8. Furthermore, if the A-type missile successfully breaks through the intercept, that the probability of successful attack is assumed to be $q = 0.5$. If the defenders are not equipped with a C-type defense system, the original defensive forces have a probability S of a successful intercept, which is assumed to be 0.3. As for the conventional weapon, the probability of successful attack is $q = 0.5$.

The attacker and defender decisions are represented as decision trees in Fig. 1. The optimal decision for the defender is simple to solve as if we know the attacker's decision.

We assume that the attacker can observe the defender's choice. If the defender does not upgrade the C-type defensive system, the attacker selects the A-type missile, and if the defender upgrades the C-type defensive system then the attacker switches to conventional weapons, then we consider that the attacker was deterred by the defender's upgrade strategy, upgrade C-type defense system is the optimal strategy. We also think that the upgrade decision is dictated by the preference of the attacker. In the

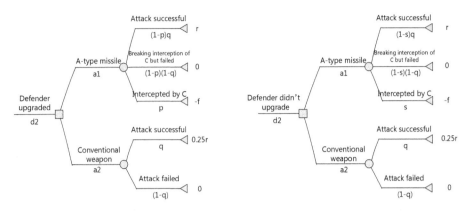

Fig. 1. The attacker decision tree if the defenders upgraded (left) and if the defender did not upgrade (right) their defenses.

following, we study the decisions of the attacker and the effect of loss aversion and likelihood insensitivity to it.

We must choose an appropriate form for the value function v and the probability weighting function π. We will choose parameterized forms to demonstrate the effects of the magnitude of loss aversion and likelihood insensitivity. We choose the simplest parametric forms of each for illustrative purposes. Tversky and Kahneman [8] introduced a value function that represents loss aversion, written as:

$$v(x)=\begin{cases} x^{\alpha} & x\geq 0 \\ -\lambda(-x)^{\beta} & x<0 \end{cases} \quad (4)$$

where $\alpha, \beta, \lambda \geq 1$. This form represents loss aversion by the increased steepness for negative values of x.

If $\alpha = \beta = 1$ then the attacker is risk neutral for prospects involving only gains and for prospects involving only losses; otherwise the value function is S shaped. This form represents no loss aversion when $\lambda = 1$, with the level of loss aversion increasing in λ. Chateauneuf et al. [12] discuss additive probability weighting function, defined by:

$$\pi(p) = \begin{cases} 1 & p = 1 \\ \kappa p + \frac{1}{2}(1 - \kappa) & 0<p<1 \\ 0 & p = 0 \end{cases} \quad (5)$$

where $0 < \kappa \leq 1$. This form can represent no probability weighting when $\kappa = 1$.

Tversky and Kahneman [8] experimentally estimated $\lambda = 2.25$. Novemsky and Kahneman [13] found individual values of λ to vary between 1 and 3, with most people's behavior corresponding to values between 2 and 3. Baillon et al. [14] finds likelihood insensitivity for the new additive weighting function equivalent to values of κ between 0.6 and 0.8. Thus, in the following the effects of loss aversion and likelihood insensitivity are tested for λ between 1 and 3 and κ between 0 and 1. The effect of adding risk attitude is tested for values of 1 (piece-wise linear value function) and 0.5

(S-shaped value function). Note that these values are estimated from experiments with a general population and are not specific to terrorist choice under uncertainty.

When the defender upgraded the C-type defensive system, a prospect-theoretic the attacker will choose the A-type missile if:

$$
\begin{aligned}
&(\pi((1-p)q) - \pi(0))v(r) + (\pi((1-p)(1-q) + (1-p)q) - \pi((1-p)q))v(0) \\
&+ (\pi(p) - \pi(0))v(-f) \geq (\pi(q) - \pi(0))v(\tfrac{1}{4}r) + (\pi((1-q)+q) - \pi(q))v(0)
\end{aligned}
\tag{6}
$$

Using the $\pi(0) = 0$ and $v(0) = 0$, we simplify it to:

$$
\pi((1-p)q)v(r) + \pi(p)v(-f) \geq \pi(q)v(\tfrac{1}{4}r)
\tag{7}
$$

Substituting the parameters of the value function, the probability weighting function and the value of p, q, r, f mentioned above ($p = 0.8$, $q = 0.5$, $r = 40$, $f = 4$), then simplifying, we obtain the following inequality:

$$
\begin{aligned}
\lambda &< \frac{(\kappa(q - pq) + \tfrac{1}{2}(1 - \kappa))r^\alpha - (\kappa q + \tfrac{1}{2}(1 - \kappa))(\tfrac{1}{4}r)^\alpha}{(\kappa p + \tfrac{1}{2}(1 - \kappa))f^\beta} \\
&= \frac{((1 - (\tfrac{1}{4})^\alpha)(\kappa q + \tfrac{1}{2}(1 - \kappa)) - \kappa pq)r^\alpha}{(\kappa p + \tfrac{1}{2}(1 - \kappa))f^\beta} \\
&= \frac{(\tfrac{1}{2}(1 - (\tfrac{1}{4})^\alpha) - \tfrac{2}{5}\kappa)40^\alpha}{(\tfrac{4}{5}\kappa + \tfrac{1}{2}(1 - \kappa))4^\beta}
\end{aligned}
\tag{8}
$$

Similarly, if the defender didn't upgrade (right-hand-side of Fig. 1), setting the parameters of the value function, the probability weighting function and the value of p, q, r, f, s mentioned   above   ($p = 0.8$, $q = 0.5$, $r = 40, f = 4, s = 0.3$),   a prospect-theoretic attacker will choose the A-type missile if:

$$
\begin{aligned}
\lambda &< \frac{(\kappa(q - sq) + \tfrac{1}{2}(1 - \kappa))r^\alpha - (\kappa q + \tfrac{1}{2}(1 - \kappa))(\tfrac{1}{4}r)^\alpha}{(\kappa s + \tfrac{1}{2}(1 - \kappa))f^\beta} \\
&= \frac{((1 - (\tfrac{1}{4})^\alpha)(\kappa q + \tfrac{1}{2}(1 - \kappa)) - \kappa sq)r^\alpha}{(\kappa s + \tfrac{1}{2}(1 - \kappa))f^\beta} \\
&= \frac{(\tfrac{1}{2}(1 - (\tfrac{1}{4})^\alpha) - \tfrac{3}{20}\kappa)40^\alpha}{(\tfrac{3}{10}\kappa + \tfrac{1}{2}(1 - \kappa))4^\beta}
\end{aligned}
\tag{9}
$$

We set $\alpha = \beta = 1$ and $\alpha = \beta = 0.5$ to represent two different kinds of risk attitude. If $\alpha = \beta = 1$, the inequalities (8) and (9) can be reduced to, respectively:

$$
\lambda < \frac{37\tfrac{1}{2} - 40\kappa}{5 + 3\kappa}
\tag{10}
$$

and

$$\lambda < \frac{75 - 30\kappa}{20 - 4\kappa} = 7.5 \tag{11}$$

When $\alpha = \beta = 0.5$, inequalities (8) and (9) can be reduced to, respectively:

$$\lambda < \frac{\sqrt{10}(\frac{5}{2} - 4\kappa)}{5 + 3\kappa} \tag{12}$$

and

$$\lambda < \frac{\sqrt{10}(5 - 3\kappa)}{10 - 4\kappa} \tag{13}$$

Figure 2 shows the different the attacker's decisions regions as a strategy plot, setting α and β, and varying λ and κ. Since $p > s$ and $\kappa > 0$, the right side of Inequality (8) is less than the right side of Inequality (9), the attacker prefers A-type missile attack to have higher loss aversion thresholds if the C-type defensive system is upgraded. This means that for the values of λ and κ which Inequality (9) does not hold, Inequality (8) also does not hold. For such values, as shown in the white regions to the right of Fig. 2, the attacker will select the conventional weapon irrespective of the defender's decision. The level of loss aversion in this area means that the possibility of successful interception will reduce the value of A-type missile attack, regardless of whether or not the defender upgraded. However, there is a tradeoff between loss aversion and likelihood insecurity at the borders of the region.

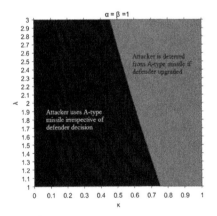
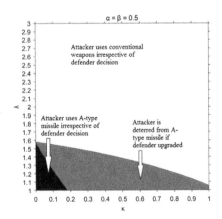

Fig. 2. Strategy plots showing the effect of loss aversion and likelihood insensitivity on the attacker's reaction to the defender's decision.

The left side of Fig. 2 shows the case of $\alpha = \beta = 1$ and the right side shows the case of $\alpha = \beta = 0.5$. For a neutral risk ($\alpha = \beta = 1$), there is no white area, so the attacker either always chooses A-type missile or is deterred by promotion and choose A-type missile only if the defender doesn't upgrade. A risk attitude involving $\alpha = \beta = 0.5$ reduces the area of the black area where the attacker always uses A-type missile and the gray area that the attacker uses A-type missile only if the defender doesn't upgrade; the white area where the attacker always chooses conventional weapons grows accordingly. In the observed regions of the experimental values of λ and κ ($2 \leq \lambda \leq 3$ and $0.6 \leq \kappa \leq 0.8$), the attacker mostly deters by promotion with the neutral risk and chose conventional weapons regardless of promotion when $\alpha = \beta = 0.5$.

Likelihood insensitivity increases the relative value of A-type missile attacks if defender did not upgrade, and the choice of conventional weapons requires more loss aversion. For values of λ and κ where inequality (8) holds, we have that (9) also holds. For such a value (represented as a black area in Fig. 2), the attacker chooses an A-type missile attack, regardless of the defender's decision. For the remaining values of λ and κ, if the defender upgraded, the attacker will select the conventional weapon, and if the defender did not upgrade, the attacker will select the A-type missile attack, so the promotion deters it from the A-type missile attack (shown as the gray area in Fig. 2).

4   Secrecy and Upgrade Randomly

The defense system is not upgraded in all locations. In this section, we consider the security measures, and only select a certain percentage of sites to upgrade. We consider the case when the scale and location of the upgrade are chosen at random and the attackers do not know whether the target is upgraded or not. Figure 3 shows the decision tree of the attacker's decision as the defender randomly upgrades the defense system with probability d.

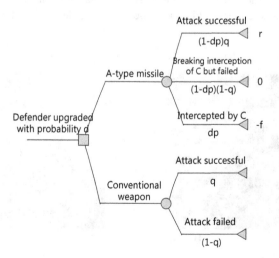

Fig. 3. The attacker decision trees if the defender upgraded with probability d

When the defender upgraded the C-type defensive system with probability d, a prospect-theoretic attacker will choose the A-type missile if:

$$
(\pi((1-dp)q) - \pi(0))v(r) + (\pi((1-dp)(1-q) + (1-dp)q) - \pi((1-dp)q))v(0)
$$
$$
+ (\pi(dp) - \pi(0))v(-f) \geq (\pi(q) - \pi(0))v(\tfrac{1}{4}r) + (\pi((1-q)+q) - \pi(q))v(0)
$$
(14)

Substituting the parameters of the value function, the probability weighting function and setting $p = 0.8, q = 0.5, r = 40, f = 4$, we obtain:

$$
\lambda < \frac{((1-(\tfrac{1}{4})^{\alpha})(\kappa q + \tfrac{1}{2}(1-\kappa)) - \kappa dpq)r^{\alpha}}{(\kappa dp + \tfrac{1}{2}(1-\kappa))f^{\beta}}
$$
$$
= \frac{(\tfrac{1}{2}(1-(\tfrac{1}{4})^{\alpha}) - \tfrac{2}{5}d\kappa)40^{\alpha}}{(\tfrac{4}{5}d\kappa + \tfrac{1}{2}(1-\kappa))4^{\beta}}
$$
(15)

We still consider the two risk attitudes of $\alpha = \beta = 1$ and $\alpha = \beta = 0.5$, and indicate the region between $d = 0$ and $d = 1$, which is called the "Waiting and see area" of the attacker.

The "Wait and see" of the attacker is shown as the gray region in Fig. 4, which is a middle area contained by two different values 0 and 1 of the parameter d in Inequality (15), which means when the attacker in this region, his decision-making is infected by the defender's value, if the point of λ and κ corresponding to the attacker is above Inequality (15), then the attacker uses conventional weapons irrespective of defender's decision, on the contrary, the attacker uses A-type missile regardless of the promotion. It is useful that the defender could choose the value of d depend on the attacker's value of λ and κ.

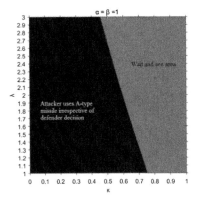 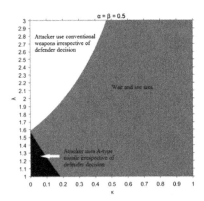

Fig. 4. Strategy plots showing the effect of loss aversion and likelihood insensitivity on the attacker's reaction to the defender decision (Defender upgrade randomly).

The case of an attacker with $\lambda = 1.4$ and $\kappa = 0.3$ is shown in Fig. 5. In this case, the defender can choose the Wait and see area under the point, and force the attacker not to use A-type missile. Thus, understanding the level of loss aversion and likelihood insensitivity of a possible attacker is important to determine the optimal level of d.

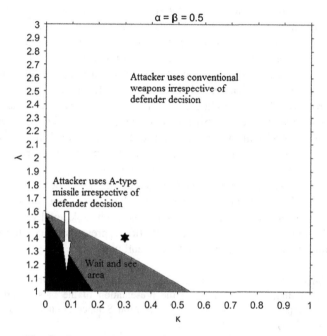

Fig. 5. Strategy plots when $\lambda = 1.4$, $\kappa = 0.3$ and d = 0.5

5   Conclusions

In this article, we have reviewed the current methods in confrontation. We chose the prospect theory to describe the loss aversion and likelihood insensitivity in the attacker's decision behavior and provide some suggestions.

We first studied the strategy of whether or not a defender upgraded into a C-type defense system, used prospect theory to build descriptive modeling of the attacker's decisions. When the attacker is at the most common level of loss aversion, likelihood insensitivity and risk attitude, he uses conventional weapons regardless of the promotion. This means that in the face of such opponents, the defender need not to upgrade the defense system. When loss aversion is low, likelihood insensitivity and the risk attitude are neutral. The Attacker uses A-type missile when the defender did not upgrade and conventional weapons if the defender upgraded. So promotion prevents such attackers from attacking to the lower consequences. Therefore, it is important to understand the level of loss aversion, likelihood insensitivity and risk attitude of the potential attackers.

The attackers do not know whether their target was upgraded and upgrading only some of the sites. This approach resulted in a large "Wait and see area", meaning that the defender could take the initiative based on the type of attacker's behavior to upgrade. This shows that understanding the level of loss aversion, likelihood insensitivity and risk attitude is important to defender's decision-making.

There are several open areas for research revealed in this work. Only one evaluation standard for the attacker and defender is considered here. Multiple objectives which represent the attacker can be included. This, however, requires the development of a descriptive model for attacker. This method requires that attributes be ordered according to their importance and considered one at a time. When considering a particular attribute, then alternatives below the expected level are removed. And finally, we get a set from which we can choose a decision that fits our preference.

Note that the results are obtained in one step. What if the attacker considers multiple sequential events in his decision, not only a sum probability? How to apply prospect theory to the decision tree with multiple sequential uncertainty nodes?

Last, prospect theory is not the only descriptive decision model. In fact, there are several alternative models of choice behavior, including rank dependent utility [15, 16] and regret theory [17, 18]. Further, there are behavioral game theory models that represent human behavior in strategic interactions, such as k-level thinking and cognitive hierarchy theory [19]. Thus, our contribution represents a first step in applying descriptive attacker models in confrontation research.

Acknowledgments. We are thankful to the Editors and the reviewers for their valuable comments and detailed suggestions to improve the presentation of the paper. Further, we also acknowledge the support in part by the National Natural Science Foundation of China under Grants No.71690233, NO.71501182.

References

1. Von Neumann, J., Morgenstern, O.: Theory of Games and Economic Behavior. Princeton University Press, Princeton (1945)
2. Parnell, G.S., Smith, C.M., Moxley, F.I.: Prospect theory intelligent adversary risk analysis: a bioterrorism risk management model. Risk Anal. **30**(1), 32–48 (2010)
3. Rios Insua, D., Rios, J., Banks, D.: Adversarial risk analysis. J. Am. Stat. Assoc. **104**(486), 841–854 (2009)
4. Simon, H.A.: A behavioral model of rational choice. Quart. J. Econ. **69**(1), 99–118 (1955)
5. Kahneman, D., Tversky, A.: Judgment under uncertainty: heuristics and biases. Science **185** (4157), 1124–1131 (1974)
6. Merrick, J., Parnell, G.S.: A comparative analysis of PRA and intelligent adversary methods for counter-terrorism risk management. Risk Anal. **31**(9), 1488–1510 (2011)
7. Kahneman, D., Tversky, A.: Prospect theory: an analysis of decision under risk. Econometrica **47**, 263–292 (1979)
8. Tversky, A., Kahneman, D.: Advances in prospect theory: cumulative representation of uncertainty. J. Risk Uncertain. **5**(4), 297–323 (1992)
9. Mazicioglu, D., Jrw, M.: Behavioral modeling of adversaries with multiple objectives in counter-terrorism. Risk Anal. (2) (2017)

10. Kujawski, E.: Accounting for terrorist behavior in allocating defensive counter-terrorism resources. Syst. Eng. **18**(4), 365–376 (2015)
11. Merrick, J.R., Leclerc, P.: Modeling adversaries in counter-terrorism decisions using prospect theory. Risk Anal. **36**(4), 681–693 (2016)
12. Chateauneuf, A., Eichberger, J., Grant, S.: Choice under uncertainty with the best and worst in mind: neo-additive capacities. J. Econ. Theory **137**, 538–567 (2007)
13. Novemsky, N., Kahneman, D.: The boundaries of loss aversion. J. Mark. Res. **42**, 119–128 (2005)
14. Baillon, A., Bleichrodt, H., Keskina, U., L'Haridonb, O., Lia, C.: Learning under Ambiguity: An Experiment Using Initial Public Offerings on a Stock Market. CREM (2014). http://EconPapers.repec.org/RePEc:tut:cremwp:201331
15. Quiggin, J.: A theory of anticipated utility. J. Econ. Behav. Organ. **3**(4), 323–343 (1982)
16. Quiggin, J.: Generalized Expected Utility Theory: The Rank-Dependent Model. Kluwer Academic Publishers, Boston (1993)
17. Bell, D.E.: Regret in decision making under uncertainty. Oper. Res. **30**(5), 960–981 (1982)
18. Loomes, G., Sugden, R.: Regret theory: an alternative theory of rational choice under uncertainty. Econ. J. **92**(4), 805–824 (1982)
19. Camerer, C.: Behavioral Game Theory: Experiments in Strategic Interaction. Princeton University Press, Princeton (2003)

A Multi-stakeholder Approach to Energy Transition Policy Formation in Jordan

Mats Danielson[1,2], Love Ekenberg[1,2(✉)],
and Nadejda Komendatova[2,3]

[1] Department of Computer and Systems Sciences, Stockholm University,
Postbox 7003, 164 07 Kista, Sweden
mats.danielson@su.se
[2] International Institute for Applied Systems Analysis, IIASA,
Schlossplatz 1, 2361 Laxenburg, Austria
{ekenberg,komendan}@iiasa.ac.at
[3] Department of Environmental Systems Science,
Institute for Environmental Decisions (ETH), Zurich, Switzerland

Abstract. We present the method used in an ongoing project in Jordan for a multi-stakeholder, multi-criteria problem of formulating a nationwide energy strategy for the country for the next decades. The Jordanian government has recognized the need for energy transition and the main goal of the energy strategy is to provide a reliable energy supply by increasing the share of local energy resources in the energy mix, while reducing dependency on imported fossil fuels, by diversifying energy resources, also including renewable energy sources, nuclear and shale oil, and by enhancing environmental protection. There were strong incentives for a collaborative approach, since the ways in which different stakeholder groups subjectively attach meanings to electricity generation technologies are recognized as important issues shaping the attainment of energy planning objectives. To understand the meaning of these constructs, we are using a multi-stakeholder multi-criteria decision analysis (MCDA) approach to elicit criteria weights and valuations.

Keywords: Energy policies · Multi-stakeholder workshops
Multi-criteria decision analysis · Surrogate criteria weights · Robustness

1 Introduction

Energy transition is a complex process which has political, social, economic and technical dimensions, requiring holistic, inclusive and comprehensive governance approach... The process of introducing energy sources and technologies can result in major socio-technical changes which might lead to frictions and conflicts. Thus, these processes will not only lead to technological changes but will also lead to socio-technological transition processes, combined with shifts in generation and distribution technologies, business models, governance structures, consumption patterns, values and world views. For sustainable implementations of these processes, new forms of governance are needed based on compromise solutions.

© Springer International Publishing AG, part of Springer Nature 2018
Y. Chen et al. (Eds.): GDN 2018, LNBIP 315, pp. 190–202, 2018.
https://doi.org/10.1007/978-3-319-92874-6_15

Energy transition could be seen as action fields or arenas where different individual or organized stakeholders are competing for legitimization of their actions and organizational survival in the future [18]. For instance, the energy sector in Jordan is well established with an existence of large providers, most often owned by the state, which generate, transmit and distribute electricity to consumers. Electricity providers, such as coal, oil and gas companies, are regarded as incumbents, i.e. actors, who have a disproportionate influence, and whose views and interests often are reflected in a dominant organization of the strategic action field, which might be entirely shaped by the worldviews, interests and positions of these incumbents. Thus, the appearance of new technologies or governance modes may heavily challenge the power distribution within the sector.

Furthermore, the dependency on imported energy sources is a heavy burden for the socio-economic and energy security of Jordan. During the last decades, energy supply to Jordan has been very volatile, also because of a number of external political shocks and setbacks. For example, an increase in the prices of crude oil, which happened during the Arab Spring in Egypt, significantly affected Jordan (being dependent on energy imports from Egypt). The interruption of Egyptian gas supply forced Jordan to switch to much more expensive heavy fuels, creating a large burden on the Jordanian national budget and significantly increased the already existing budget deficit. Also, to handle the difference between imported energy and affordability in the local market, the Jordanian government needed to heavily subsidize energy imports, which further increased its national deficit.

Several reports on energy transition in Jordan have been written, with the focus on economic and technological factors. Following evidence as well as national and international advice, Jordan has been developing a legal and regulatory framework to attract investments in renewable energy expansion but also in new technologies such as shale oil and nuclear power. However, the discussions about an energy transition and a transformation of the Jordanian society, which might be caused by large-scale deployment of new technology, as well as socio-economic consequences of the transformation of the energy system, have been limited. Furthermore, the process of learning from other regions and from technology transfer, which goes beyond single projects, but includes regional models of energy transitions and transformation of society, should be considered with caution. There are several examples and good practices in Europe, such as "Energiewende" in Germany or energy transition through climate and energy models in Austria. However, a plan for an energy transition in the MENA region should consider completely different energy market structures, stakeholders' networks and societal aspirations regarding energy, climate and environmental policies [22].

MCDM methods have been used during the last decades to select between different energy system solutions, most often on a regional scale, for example [2, 24, 34], or smaller scale [29], or for non-specific discussions on energy system solutions [25] or policies [16]. Some have had a national scope targeting a specific technology, for example [36]. This current project, however, deals with a national energy policy based on input information from large sets of stakeholders. The required methodology to deal with it is a multi-stakeholder, multi-criteria decision analysis method suitable for discussions and negotiations in different settings and with different background data.

In this article, we discuss, an ongoing project for the multi-stakeholder, multi-criteria problem of formulating an energy strategy for Jordan for the next decades.

2   Problem Setup

This section describes the process of selecting a relevant energy policy and makes this process more transparent. It also identifies a set of optimal solutions out of a set of technologies of the prevailing realistic options.

2.1   Identification of Available Technologies

The collection of data for establishing the performance characteristics for each technology was based on different sources and methods encompassing both quantitative and qualitative data. Primary quantitative data sources involved remote sensing data and Geographical Information System (GIS) maps as well as data available from national statistics databases. Secondary quantitative data sources included a total of more than 200 regionally specific and international peer-reviewed articles, official policy reports, industry reports, Environmental and Social Impact Assessments (ESIAs), and real project case studies. Additionally, experts were surveyed to obtain qualitative indicators on criteria where no quantitative data could be found or developed, such as the perceptions of capacity of national authorities to control the risks involved. A purposeful sampling was applied in order to consult a balanced diversity of experts from different fields of expertise and roles in society. The identification and selection of individuals was influenced by practical considerations, such as the availability, willingness to participate, or opportunities that emerged during the research process [26]. Overall, 52 experts were asked to take part in the survey of whom 31 responded. The identification of the technologies resulted in the following set of options:

- **Utility-Scale Photovoltaic** (PV), which uses direct and diffused solar radiation and converts it into electricity by using a photovoltaic effect.
- **Concentrated Solar Power** (CSP), which, with the help of different kinds of mirrors, concentrates solar radiation onto a receiver and then converts it into thermal energy inside the receiver. Then thermal energy is transformed into mechanical energy with the help of a steam turbine and converted to electricity with the help of a generator.
- **Onshore Wind**, which with the aid of wind turbines harnesses kinetic energy of the wind and converts it first into mechanical energy and then electricity.
- **Utility Hydro-Electric** uses water to turn a turbine that provides mechanical energy and drivers a generator.
- **Nuclear Power**, which uses the thermal energy released by uranium fission reactions.
- **Lignite Coal**, when the coal-fired power plant converts the chemical energy from coal into heat in the process of combustion. The heat is then used to generate steam which drives a turbine to produce electricity.

- **Natural Gas**, when kinetic energy from the motion of flowing gas is utilized to generate electricity with the help of a gas turbine.
- **Heavy Fuel Oil**, when the oil-fired power plant uses the chemical energy of oil to generate electricity with the help of different kinds of steam systems.

These technologies were identified before consultations with the stakeholder groups and were presented for comments, after which the stakeholders added shale oil to the set of options.

2.2 Criteria Selection

All technologies were assessed against a set of 11 criteria, with a corresponding total of 20 indicators. Out of these, 9 indicators are quantitative and 11 are qualitative. The criteria were selected in a threefold, iterative process. The first step of the selection process was based on an extensive literature review of scientific publications that developed criteria relevant to assessing the performance of energy systems and electricity technologies (e.g., [1, 11, 12, 19–21, 23, 35]). Thereafter, the national policy framework was supplemented with a criterion set with nationally relevant development criteria. Each criterion was then evaluated according to its relevance for the decision-making problem at hand ("high", "medium", or "low"). This process included several interactions and iterations, through which the number of criteria was eventually narrowed down from 32 to the final set of 11 criteria in a three-level criteria with tree comprising a total of 24 sub-criteria, see Fig. 1.

2.3 Stakeholder Groups

At the core of the study was a series of seven one-day stakeholder workshops. Each of the first six workshops included groups of stakeholders from the same backgrounds, whereas the participants in the final workshop comprised a heterogeneous stakeholder group to which an equal number of previous participants from each of the previous workshops (i.e. stakeholder groups) were invited.

In line with different scholars [3, 30, 31, 34], who recommend the inclusion of political, economic, scientific, and socio-cultural actors in electricity planning, six stakeholder groups of different backgrounds were selected to participate in this research. The participants were identified based on a comprehensive stakeholder analysis and according to their positions and interest in Jordan's energy decision-making and also based on the extent to which they are impacted by electricity installations. The stakeholder analysis was conducted by the research team in cooperation with the local partners. In the first step, broad stakeholder categories, for example "Policy-makers", "Young Leaders", etc. in line with the above mentioned different backgrounds were established. In the second step, these categories were broken down into more concrete sub-groups of these categories, e.g., for the category "Finance and Industry", small and medium-sized enterprises or national banks.

In the final step, the representatives of the sub-groups were determined. As a result, the following stakeholder groups were identified.

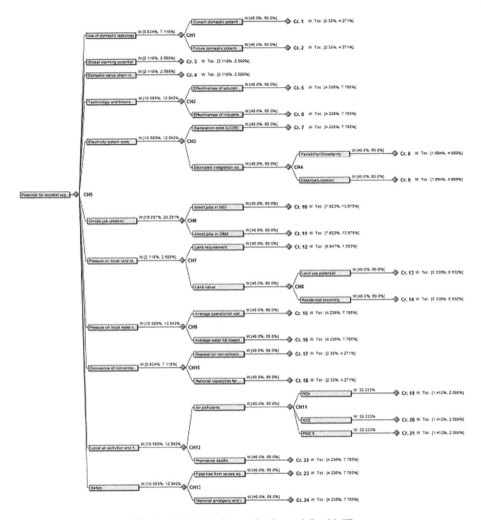

Fig. 1. Final criteria tree in the tool DecideIT.

- **Policy-makers:** stakeholders who are directly involved in electricity planning including generation and distribution;
- **Finance and Industry:** stakeholders who are characterized by high electrical power end-use and are directly involved in the implementation and financing of power generation capacities;
- **Academia:** stakeholders who are scientifically interested and involved in the research and development of electricity systems, e.g., universities, research institutions, and think tanks;
- **Young Leaders:** stakeholders who can be regarded as future decision-makers or opinion-leaders and have a strong interest in national energy planning due to their professional background, public engagement, and networks;

- **National NGOs:** stakeholders who have a strong interest in national energy planning and are involved in national NGOs working on environmental protection, social justice, and human development;
- **Local Communities:** stakeholders who live in close proximity to electricity generation technologies and are thus directly affected by national electricity planning;

The involvement of different stakeholder groups, in an intensive, discussion-oriented, and participatory process, allowed a wide array of multidimensional perspectives on Jordan's energy future to be elicited. During the workshops, attention had to be focused on creating a climate that welcomed discussion, where different stakeholder views were respected and equally validated, while at the same time room for mutual learning and new information was provided. This was in particular the case for the final workshop where equally legitimate opinions and perspectives as well as mutual learning experiences had to be safeguarded by the moderators in order to allow for a balanced representation of all stakeholder groups during the often heated debates among participants.

During the stakeholder workshops the participants were given 45 min to develop their individual vision on what they, as representatives of their specific stakeholder group, would like Jordan to be in the year 2050. The participants were provided with a set of cards and asked to write down either a short sentence or an attribute for their vision in three areas: the economy, the society, and the environment. Then, all cards were discussed on flipcharts and clustered in common themes, where agreement was reached, as well as where perspectives diverged. The aim here was not to be comprehensive but rather to identify the top priorities with respect to how different stakeholder groups imagine a desirable future for Jordan.

Following the vision development phase, the participants were given 60 min to express their aspirations and concerns in regard to the question of how the deployment of new electricity generation technologies in Jordan could enable or hamper the fulfilment of their vision financially, socially, and environmentally. To facilitate this task, short vignettes for each technology were distributed to provide as far as possible unbiased, non-technical information on the basic functions and performance of the technologies under examination. After discussing the specifics of the different technologies, the participants were again provided with a set of cards and asked to formulate their thoughts as representatives of their specific stakeholder groups. Afterwards, all cards were clustered around the main issues that emerged during the vision development and discussed openly.

3 Criteria Ranking

Simos proposed a procedure, using a set of cards, for determining numerical values for criteria weights suitable for negotiation settings [32, 33]. In its standard form, a group of decision-makers are provided with a set of coloured cards with the criteria names written on them. Furthermore, the decision-makers are provided with a set of white (blank) cards. Thereafter, the non-blank cards are ranked from the least important to the most important, where criteria of equal importance are grouped together. Furthermore, the decision-makers are asked to place the white cards in between the coloured cards to express preference strengths. Figueira and Roy [17] have suggested a modified version, where the decision-makers state how many times the most important criterion or criteria group is compared to the least important. A variation of the Simos method was used in this project for elicitation purposes. The card ranking part was employed as the original but the evaluation part differs significantly from the Simos method. The criteria were at this point well-known by the participants from the previous sections of the workshops.

Each criterion was written on a coloured card and arranged horizontally on a table. Then each of the participants successively ranked the cards from the least important to the most important by moving the cards to a vertical arrangement, where the highest ranked criteria was put on top and so forth. If two criteria were considered to be of equal importance, they were put on the same level. This process lasted for four rounds, where the number of moves for each round was 8, 5, 3 and 2 respectively. Furthermore, the first and third round was concluded by an open discussion before the following round commenced.

The ranking procedure lasted for 120 min or until a final ranking was obtained that the participants found acceptable. Needless to say, the decreasing number can be disputed and is a weak point of the method (and thus an open research question needs to be addressed in subsequent projects), since it encourages the participants to act strategically in relation to the information they received during the process. So when applying this method, the potential conflicts must be lifted and handled. In some cases, working with a set of final ranking in the evaluations, showed whether or not the differences are of importance.

When this first ordinal ranking was finalised, the participants were asked to introduce preference strengths in the ranking by introducing the blank cards during three additional rounds (with 3, 2 and 1 moves respectively). The number of white cards (i.e. the strength of the rankings between criteria) was also given a verbal interpretation as shown in Table 1.

Table 1. Verbal interpretation of card placements

Equal level of cards	Equally important
No blank card	Slightly more important
One blank card	More important (clearly more important)
Two blank cards	Much more important
Three blank cards	Very much more important

The final rankings of the six workshops were handed to the representatives of each stakeholder group during the final workshop two months later, when the exercise was repeated also for this compiled cross-sectional multi-stakeholder group. At the final workshop, they could present each ranking and its rationales to the other participants during an introductory presentation round.

4   Selection of Analysis Method

One well-known class of multi-criteria decision analytic methods is the SMART family, where [13–15] proposed a method to assess criteria weights. The criteria are ranked and then 10 points are assigned to the weight of the least important criterion (w_N). Then, the remaining weights (w_{N-1} through w_1) are given points according to the decision-maker's preferences. The overall value $E(a_j)$ of an alternative a_j is then the weighted average of the values v_{ij} associated with a_j (Eq. 1):

$$E(a_j) = \frac{\sum_{i=1}^{N} w_i v_{ij}}{\sum_{i=1}^{N} w_i} \tag{1}$$

The most utilised processes for converting ordinal input to cardinal use automated procedures and yield exact numeric weights. For instance, [13] proposed the SMARTER method for eliciting ordinal information on importance before converting it to numbers, thus relaxing information input demands on the decision-maker. An initial analysis is carried out where the weights are ordered, such as $w_1 > w_2 > \ldots > w_N$, and then subsequently transformed to numerical weights using ROC weights after which SMARTER continues in the same manner as the ordinary SMART method.

The best known ratio scoring method is the Analytic Hierarchy Process (AHP). The basic idea in AHP [27, 28] is to evaluate a set of alternatives under a criteria tree by pairwise comparisons. The process requires the same pairwise comparisons regardless of scale type. For each criterion, the decision-maker should first find the ordering of the alternatives from best to worst. Next, he or she should find the strength of the ordering by considering pairwise ratios (pairwise relations) between the alternatives using the integers 1, 3, 5, 7, and 9 to express their relative strengths, indicating that one alternative is equally good as another (strength = 1) or three, five, seven, or nine times as good. It is also allowed to use the even integers 2, 4, 6, and 8 as intermediate values, but using only odd integers is more common.

As discussed in [5], a viable alternative to these, when cardinal information is present, is the Cardinal Ranking (CAR) method and it has been demonstrated that the latter is more robust and efficient than the methods from the SMART family and AHP. Providing only ordinal rankings of criteria seems to avoid some of the difficulties associated with the elicitation of exact numbers. It puts fewer demands on decision-makers and is thus, in a sense, effort-saving. Furthermore, there are techniques for handling ordinal rankings with various successes. A limitation of this approach is that decision-makers, not least in multi-stakeholder settings, usually have more knowledge of the decision situation than a pure criteria ordering is able to capture, often in the sense that they have an idea regarding the importance of relation information containing strengths. In such cases, the ordinal rankings are often an unnecessarily weak representation, leading to a need for extending the methods to accommodate information regarding relational strengths as well, while still preserving the property of being less demanding and more practically useful than other types of methods such as SMARTS or AHP. In line with the results in [9], the CAR method was selected for the evaluation phase in the project.

5   Evaluations

The analytical part of the evaluation in the project consists of translating the rankings to surrogate weights, evaluating them by applying the CAR method, and then using these values in the software DecideIT which is designed for solving this type of problem under uncertainty. Thereby, the information loss is limited [8]. The idea is the following:

- Assign an ordinal number to each importance scale position, starting with the most important position as number 1.
- Let the total number of importance scale positions be Q. Each criterion i has the position $p(i) \in \{1, ..., Q\}$ on this importance scale, such that for every two adjacent criteria c_i and c_{i+1}, whenever $c_i >_{s_i} c_{i+1}$, $s_i = |p(i + 1) - p(i)|$. The position $p(i)$ then denotes the importance as stated by the decision-maker. Thus, Q is equal to $\sum s_i + 1$, where $i = 1, ..., N - 1$ for N criteria.

In [9, 10], several cardinal weight methods are derived, discussed, and evaluated. The best method for cardinal rankings with properties similar to Simos cards was shown to be CSR weights, expressed as

$$w_i^{CSR} = \frac{1/p(i) + \frac{Q + 1 - p(i)}{Q}}{\sum_{j=1}^{N}\left(1/p(j) + \frac{Q + 1 - p(j)}{Q}\right)} \tag{2}$$

which are consequently employed in this study. Based on the weightings of each stakeholder group, expressed as CSR weights, and observations made during the workshops, the analysis of potential conflict lines and commonalities between the different stakeholder preferences was facilitated through negotiation.

5.1   Encoding of Criteria Weights

As mentioned above, one of the problems with most models for criteria ranking is that numerically precise information is seldom available. This is partially solved by introducing surrogate weights in the way that was done before, but this is only a part of the solution since the elicitation can still be uncertain and the surrogate weights might not be a totally adequate representation of the preferences involved, which is of course a risk with all kinds of aggregations. To allow for a more thorough robustness analysis, we also introduce intervals around the derived weights as well as around the values of the technology options. Thus, in this elicitation problem, the possibly incomplete information is handled by allowing the use of ranges of possible values (cf., e.g., [4, 6, 7]).

There are thus several approaches to elicitation in MCDM problems and one partitioning of the methods into categories is how they handle imprecision in weights (or values).

- Weights (or values) can only be estimated as fixed numbers.
- Weights (or values) can be estimated as comparative statements converted into fixed numbers representing the relations between the weights.

- Weights (or values) can be estimated as comparative statements converted into inequalities between interval-valued variables.
- Weights (or values) can be estimated as interval statements.

Needless to say, there are advantages and disadvantages of the different methods. Methods based on categories 1 and 2 yield computationally simpler evaluations because of the weights and values being numbers, while categories 3 and 4 yield systems of constraints in the form of equations and inequalities that need to be solved using optimisation techniques. If the expressive power of the analysis method only permits fixed numbers (category 1), we usually get a limited model that might affect the decision quality severely. If intervals are allowed (categories 3 and 4), imprecision is normally handled by allowing variables, where each y_i is interpreted as an interval such that $w_i \in [y_i - a_i, \ y_i + b_i]$, where $0 < a_i \leq 1$ and $0 < b_i \leq 1$ are proportional imprecision constants. Similarly, comparative statements are represented as $w_i \geq w_j$. However, we might encounter an unnecessary information loss using only an ordinal ranking. When using both intervals and ordinal information, we obtain some rather elaborate computational problems. Despite the fact that they can be solved, when sufficiently restricting the statements involved (cf. [7]), there is still a problem with user acceptance and these methods have turned out to be perceived as too difficult to accept by many decision-makers. Expressive power in the form of intervals and comparative statements leads to complex computations and loss of transparency on the part of the user. This should be kept in mind here as always when working with aggregation methods of whatever kind.

5.2   Results from the Final Workshop

The performance of different electricity generation technologies was estimated based on a large expert survey. Together with the surrogate weights, they provided the decision base for the multi-criteria analysis. Using the aggregation principle in (2), the multiple criteria and stakeholder preferences could be combined with the valuation of the different technology options under the criteria surrogate weights.

The results of the evaluations are (i) a detailed analysis of the performance of each technology compared with the other technologies, and (ii) a sensitivity analysis to test the robustness of the result. Figure 2 shows part of the results of the final workshop.

In the figure, it can be seen that alternative 1 (Utility-Scale Photovoltaic) is the preferred alternative, meaning that solar radiation converted into electricity by the photovoltaic effect was the collective stakeholders' preference. As the runner-up, alternative 2 (Concentrated Solar Power) was selected, which concentrates solar radiation onto a receiver and then converts it into thermal energy. Especially the preference of alternative 1 was very pronounced in the standings after the final stakeholder summit. Thus, it became the recommendation from the summit (the final workshop).

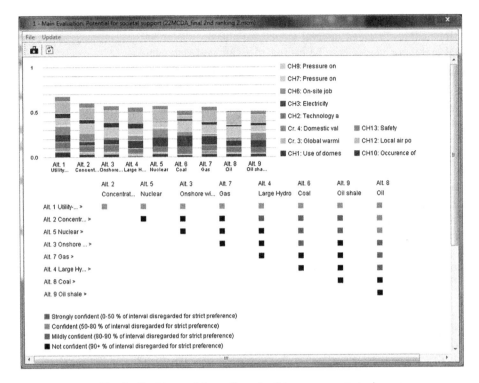

Fig. 2. Rankings of the options for future energy strategies

6   Concluding Remarks

We have presented the method of an ongoing project in Jordan for a multi-stakeholder, multi-criteria problem of formulating an energy strategy for Jordan for the next decades. We used a multi-stakeholder MCDM approach with ordinal or imprecise importance information and suggested a method for how to incorporate various stakeholders' views on energy technologies and their valuation under several criteria. The implementation of MCDM in Jordan was based first on stakeholder workshops with each of a set of groups of stakeholders and then also applied within a final concluding workshop with mixed groups of stakeholders. Our experience of the implementation of the MCDM methodology showed that the local process in Jordan could benefit from a series of workshops with the same mixed group of stakeholders. Such a process would also contribute to the identification of compromise solutions and the facilitation of discussions among stakeholders with different views and perceptions on the importance of different technology relevant criteria. We also followed up with a survey in which stakeholders were asked to rank their results again, but this time individually and not as a group, and to assess their degree of satisfaction with the results from the final workshop. This current article deals with the methodological issues in the project and we have thus omitted a discussion on the final outcome of this analysis, but this will be the subject of a forthcoming article.

Acknowledgements. This research was partly funded by the project Middle East North Africa Sustainable Electricity Trajectories (MENA-SELECT) and partly by the strategic grants from the Swedish government within ICT – The Next Generation.

References

1. Afgan, N.H., Carvalho, M.G.: Multi-criteria assessment of new and renewable energy power plants. Energy **27**, 739–755 (2002)
2. Beccali, M., Cellura, M., Mistretta, M.: Decision-making in energy planning. Application of the Electre method at regional level for the diffusion of renewable energy technology. Renew. Energy **28**, 2063–2087 (2003)
3. Bidwell, D.: Thinking through participation in renewable energy decisions. Nat. Energy **1**, 16051 (2016)
4. Danielson, M., Ekenberg, L.: A framework for analysing decisions under risk. Eur. J. Oper. Res. **104**(3), 474–484 (1998)
5. Danielson, M., Ekenberg, L.: A robustness study of state-of-the-art surrogate weights for MCDM. Group Decis. Negot. **7** (2016). https://doi.org/10.1007/s10726–016-9494-6
6. Danielson, M., Ekenberg, L., He, Y.: Augmenting ordinal methods of attribute weight approximation. Decis. Anal. **11**(1), 21–26 (2014)
7. Danielson, M., Ekenberg, L.: Computing upper and lower bounds in interval decision trees. Eur. J. Oper. Res. **181**(2), 808–816 (2007)
8. Danielson, M., Ekenberg, L.: Rank ordering methods for multi-criteria decisions. In: Zaraté, P., Kersten, G.E., Hernández, J.E. (eds.) GDN 2014. LNBIP, vol. 180, pp. 128–135. Springer, Cham (2014). https://doi.org/10.1007/978-3-319-07179-4_14
9. Danielson, M., Ekenberg, L.: The CAR method for using preference strength in multi-criteria decision making. Group Decis. Negot. **25**(4), 775–797 (2016). https://doi.org/10.1007/s10726-015-9460-8
10. Danielson, M., Ekenberg, L.: Trade-offs for ordinal ranking methods in multi-criteria decisions. In: Bajwa, D., Koeszegi, S.T., Vetschera, R. (eds.) GDN 2016. LNBIP, vol. 274, pp. 16–27. Springer, Cham (2017). https://doi.org/10.1007/978-3-319-52624-9_2
11. Del Rio, P., Burguillo, M.: Assessing the impact of renewable energy deployment on local sustainability: towards a theoretical framework. Renew. Sustain. Energy Rev. **12**, 1325–1344 (2008)
12. Del Rio, P., Burguillo, M.: An empirical analysis of the impact of renewable energy deployment on local sustainability. Renew. Sustain. Energy Rev. **13**, 1314–1325 (2009)
13. Edwards, W., Barron, F.: SMARTS and SMARTER: improved simple methods for multiattribute utility measurement. Organ. Behav. Hum. Decis. Process. **60**, 306–325 (1994)
14. Edwards, W.: How to use multiattribute utility measurement for social decisionmaking. IEEE Trans. Syst. Man Cybern. **7**(5), 326–340 (1977)
15. Edwards, W.: Social utilities. Eng. Econ. Summer Symp. Ser. **6**, 119–129 (1971)
16. Erol, Ö., Kilkis, B.: An energy source policy assessment using analytical hierarchy process. Energy Convers. Manag. **63**, 245–252 (2012)
17. Figueira, J., Roy, B.: Determining the weights of criteria in the ELECTRE type methods with a revised Simos' procedure. Eur. J. Oper. Res. **139**, 317–326 (2002)
18. Fligstein, N., McAdam, D.: A Theory of Fields. Oxford University Press, Oxford (2012)
19. Grafakos, S., Flamos, A., Enseñado, E.M.: Preferences matter: a constructive approach to incorporating local stakeholders' preferences in the sustainability evaluation of energy technologies. Sustainability **7**, 10922–10960 (2015)

20. Grafakos, S., Gianoli, A., Tsatso, A.: Towards the development of an integrated sustainability and resilience benefits assessment framework of urban green growth interventions. Sustainability **8**, 461 (2016)
21. Hirschberg, S., Burgherr, P., Spiekerman, G., Dones, R.: Severe accidents in the Energy Sector. PSI Report No. 98-16, Villigen (1998)
22. Komendantova, N., Irshaid, J., Marashdeh, L., Al-Salaymeh, A., Ekenberg, L., Linnerooth-Bayer, J.: Country Fact Sheet Jordan: Energy and Development at a Glance 2017: Background Paper. Middle East North Africa Sustainable Electricity Trajectories (MENA-SELECT) Project Funded by the Federal Ministry for Economic Cooperation and Development (BMZ), 65 pp. (2017)
23. Kowalski, K., Stagl, S., Madlener, R., Omann, I.: Sustainable energy futures: Methodological challenges in combining scenarios and participatory multi-criteria analysis. Eur. J. Oper. Res. **197**, 1063–1074 (2009)
24. Latinopoulos, D., Kechagia, K.: A GIS-based multi-criteria evaluation for wind farm site selection. A regional scale application in Greece. Renew. Energy **78**, 550–560 (2015)
25. Oberschmidt, J., Geldermann, J., Ludwig, J., Schmehl, M.: Modified PROMETHEE approach for assessing energy technologies. Int. J. Energy Sect. Manag. **4**, 183–212 (2010)
26. Palinkas, L.A., Horwitz, S.M., Green, C.A., Wisdom, J.P., Duan, N., Hoagwood, K.: Purposeful sampling for qualitative data collection and analysis in mixed method implementation research. Adm. Policy Ment. Health Ment. Health Serv. Res. **42**, 533–544 (2015)
27. Saaty, T.L.: A scaling method for priorities in hierarchical structures. J. Math. Psychol. **15**, 234–281 (1977)
28. Saaty, T.L.: The Analytic Hierarchy Process. McGraw-Hill, New York (1980)
29. Simos, J.: Evaluer l'impact sur l'environnement: Une approche originale par l'analyse multicriteere et la negociation. Presses Polytechniques et Universitaires Romandes, Lausanne (1990)
30. Simos, J.: L'evaluation environnementale: Un processus cognitif neegociee. Theese de doctorat, DGF-EPFL, Lausanne (1990)
31. Schweizer, P.-J., Renn, O., Köck, W., Bovet, J., Benighaus, C., Scheel, O., Schröter, R.: Public participation for infrastructure planning in the context of the German "Energiewende". Utilities Policy (2014, in press)
32. Sánchez-Lozano, J.M., Henggeler Antunes, C., García-Cascales, M.S., Dias, L.C.: GIS-based photovoltaic solar farms site selection using ELECTRE-TRI: evaluating the case for Torre Pacheco, Murcia, Southeast of Spain. Renew. Energy **66**, 478–494 (2014)
33. Shortall, R., Davidsdottir, B., Axelsson, G.: Development of a sustainability assessment framework for geothermal energy projects. Energy. Sustain. Dev. **27**, 28–45 (2015)
34. Tsoutsos, T., Drandaki, M., Frantzeskaki, N., Iosifidis, E., Kiosses, I.: Sustainable energy planning by using multi-criteria analysis application in the island of Crete. Energy Policy **37**, 1587–1600 (2009)
35. Wang, J.-J., Jing, Y.-Y., Zhang, C.-F., Zhao, J.-H.: Review on multi-criteria decision analysis aid in sustainable energy decision-making. Renew. Sustain. Energy Rev. **13**, 2263–2278 (2009)
36. Zangeneh, A., Jadid, S., Rahimi-Kian, A.: A hierarchical decision making model for the prioritization of distributed generation technologies: a case study for Iran. Energy Policy **37**, 5752–5763 (2009)

Author Index

Ahmed, Waqas   154
Ali, Sharafat   154, 167

Capolongo, Stefano   97
Carreras, Ashley L.   113
Chen, Ziyi   179

Danielson, Mats   190
de Almeida, Adiel Teixeira   56, 97
Dell'Ovo, Marta   97
Dou, Yajie   33, 179

Ekenberg, Love   190

Flegentova, Ekaterina   139
Frej, Eduarda Asfora   56, 97

Ge, Bingfeng   179

He, Shawei   139

Jiang, Jiang   33

Kaur, Parmjit   113
Kersten, Gregory E.   43, 70, 82
Komendatova, Nadejda   190

Liao, Huchang   3

Mi, Xiaomei   3
Morais, Danielle Costa   97

Oppio, Alessandra   97

Roselli, Lucia Reis Peixoto   56
Roszkowska, Ewa   43, 82

Song, Wen   14

Tan, Qingmei   154
Tan, Yuejin   33, 179
Theodora, Michelle   167

Vahidov, Rustam   127

Wachowicz, Tomasz   43, 82
Wan, Chunqi   179

Xu, Haiyan   167
Xu, Peng   167
Xu, Xiangqian   33

Yu, Bo   70

Zhou, Zhexuan   33
Zhu, Bing   139
Zhu, Jianjun   14

Printed in the United States
By Bookmasters